The Role of Values in Sustainability Transition: The Case of Chinese Ecological Agriculture

THE ROLE OF VALUES IN SUSTAINABILITY TRANSITION

THE CASE OF CHINESE ECOLOGICAL AGRICULTURE

YINGJIE WU

Westphalia Press
An Imprint of the Policy Studies Organization
Washington, DC
2024

The Role of Values in Sustainability Transition:
The Case of Chinese Ecological Agriculture

Westphalia Press
An imprint of Policy Studies Organization
1367 Connecticut Avenue NW
Washington, D.C. 20036
info@ipsonet.org

ISBN: 978-1-63723-661-1

Cover and interior design by Jeffrey Barnes
jbarnesbook.design

Daniel Gutierrez-Sandoval, Executive Director
PSO and Westphalia Press

Updated material and comments on this edition
can be found at the Westphalia Press website:
www.westphaliapress.org

Acknowledgements

I would like to express my gratitude to my two supervisors, Prof. Dr. Claudia Bieling and Dr. Uta Eser, for their invaluable input and constructive advice, which greatly improved the quality of my thesis. Particularly, Claudia Bieling guided me from the beginning of the thesis and supported me when I encountered challenges in my research.

My sincere appreciation also extends to Mrs. Julia Rietze for simplifying the administrative work, not only during the thesis period but also when I was a student assistant. Additionally, I am grateful to my sponsor, Dr. Teresa Galinsky, for her altruistic financial support during the onset of the pandemic in 2020, which allowed me to focus on my studies without worrying about making ends meet.

I am deeply humbled to have had the support of my supervisors and numerous friends who offered assistance at various stages of my thesis and during the challenging moments of my life as a master's student. I extend my sincere gratitude to all of you.

Disclaimer: The author wishes to acknowledge the use of gender-neutral language throughout this publication. For simplicity of writing, the term "man" is employed inclusively to represent humanity as a whole and is not intended to exclude any genders.

Table of Contents

Foreword: An Accomplished Study

By Paul Rich

AB EDM, Harvard University
PhD, University of Western Australia
Fellow of Royal Historical Society
President of Policy Studies Organization

Yingjie Wu has made an important contribution bridging the gap in communication between Chinese and Western scholarship by his activities as a senior editor for the Policy Studies Organization and its Westphalia Press. This new and significant study of his is exemplary. His apt editorship of so many journals and books has been a testimony both to his bibliographical ability and to the value of introducing the two worlds to each other. We sorely need better communication between those working on similar topics unknowingly.

Chinese academics are making significant contributions to academic fields and perspectives, and often come up with solutions that are original. For instance, China's current developments in artificial intelligence travel different roads from those in Europe and America. We need more people like him who can stretch boundaries and compare resources. These networks are vital for the rapid dissemination of knowledge and collaborative problem-solving. Use of Chinese sources is critical to numerous subjects.

Additionally, geopolitical tensions can be eased by the conversations between what need not be two separate academies. His ability to do research in the literature of several societies is what we need to have and it is regrettable that the lack of more bilingual scholars sadly continues to create a gulf.

Sustainability, at its most basic, is about the crises of the present and our fears of compromising the future. As this study suggests, the need for understanding its dimensions and demands are complex and require the balance of numerous factors. This essay explores some of these essential values and incorporates both Chinese and Western views successfully.

The subject certainly requires more study than it has received. The fundamental values it includes are only now getting the attention they deserve. Without more discussion, the missing transparency is what the stakeholders, including consumers, investors, and community members are now demanding. This work supports intelligent decision-making. Studies like this increase credibility and encourage greater engagement in sustainability efforts. Moreover, such inclusivity goes to wider debate about the diverse values and perspectives of all stakeholders. These values are particularly important in addressing in the future the social dimensions of sustainability and intergenerational equity as a principle that advocates for fairness between current and future generations. Intelligent examination of these values calls for prudent use of resources to ensure that future generations inherit a world that is not depleted of its wealth and health.

Scholarship at its best, as this is, probes the significance of what we often take for granted. To move forward, studies like this should be a priority. The reader will quickly see the implications of Wu's analysis.

List of abbreviations, Tables, and Figures

Abbreviations

CEA= Chinese ecological agriculture

ECC= Ecological Civilisation Construction

IPBES= Intergovernmental Science-Policy Platform on Biodiversity and Ecosystem Services

TMT= Two Mountains Theory

Figures

Tables

Glossary

Anthropogenic assets: Anthropogenic assets refer to built-up infrastructures, facilities, technologies, knowledge, etc., that originate from human activities. Their interaction with nature to produce benefits is used as a method to achieve a good quality of life (Díaz et al., 2015).

Chinese ecological agriculture (CEA): The concept of Chinese ecological agriculture was proposed by Chinese scholar Qianji Ye in 1981 (Ye, 1982). It draws on the theory of ecological agriculture, follows the principles of ecology, and combines the modern technology with traditional agricultural practices to form a sustainable agricultural system that is tailored according to different contexts. It aims to improve people's quality of life, address the dilemma between man and the environment, and thus achieve economic, social, and ecological benefits.

Ecological agriculture: There are various definitions of ecological agriculture. This study draws on the concept of ecological agriculture defined by M. Kiley-Worthington as the establishment and maintenance of a small farming system that is self-sustaining with low input, economically viable, that maximises production without causing big changes to the environment, and that is aesthetically and ethically acceptable (Kiley-Worthington, 1981).

Ecological Civilisation Construction (ECC): Ecological Civilisation Construction is a concept proposed by the Chinese government in 2012. It aims at building a resource-saving and environmentally-friendly society that maintains the carrying capacity of resources and the environment, follows the laws of nature, and contributes to the sustainable development (Gu, Hu, & Zhou, 2013).

IPBES framework: This conceptual framework is developed by Intergovernmental Science-Policy Platform on Biodiversity and Ecosystem Services (IPBES) that summarizes the relationships between man and nature (Díaz et al., 2015). Nature, nature's benefits to people, and good quality of life are three key elements of this framework.

Sustainability transition: Sustainability transitions are radical and long-term processes taking place in social-technical systems that consists of actors, networks, institutions and knowledge (Markard, Raven, & Truffer, 2012). Sector of agriculture is one example of the social-technical systems.

The purpose of such processes is to transition into a sustainable mode of production and consumption, therefore contributing to a broader sustainable society and way of living.

Two Mountains Theory (TMT): It is a theory as well as a political slogan proposed by President Xi in 2005. It explains that clear waters and green mountains are as valuable as mountains of gold and silver. The two mountains, i.e., the green mountains and the gold and silver mountains, refer to environment and economy growth respectively. It is a dynamic process that addresses the relationship between man and nature. The dialectical unity in this theory lies in three stages: exploiting nature for economic growth, a stage in which the sustainability is not considered; keeping the economic growth while protecting the nature, a stage in which the environmental issues deteriorate and man starts to realize the importance of protecting nature; in the third stage, man realizes that to keep the two mountains, it is necessary to change the development into a sustainable mode. When economic development is derived sustainably from natural resources, the clear waters and green mountains can continuously provide mountains of gold and silver—and they are the mountains of gold and silver (Xi, 2007).

Sancai: The concept of Sancai is a traditional agricultural wisdom, which refers to equal treatment of the Heaven, Earth, and people (G. Li, 2009). In this view, Heaven refers to the climate and Earth refers to the land—together they form the nature, whereas man is the subject. Man is neither the slave nor the master of nature but a steward in the interaction with nature.

Framework method: Framework method is a qualitative analysis method developed by two researchers, Jane Ritchie and Liz Spencer in late 1980s (Ritchie & Lewis, 2003). It is a matrix-based method for analysing qualitative data. The name Framework comes from "thematic framework" which is used to organize data and cluster them according to codes and categories.

Analytical framework: In Framework method, an analytical framework consists of the codes and categories developed by researchers (Ritchie & Lewis, 2003). It is helpful for organizing and summarizing the original data in a list format to guide data interpretation in a later stage.

Open coding: Open coding is an interpretive process where data are broken down into sections and each section is labelled conceptually by assigning a code (Corbin & Strauss, 1990). A code can be a word or phrases that extract the most important information. Conceptually related codes are grouped together into categories for further analysis.

Charting: In the Framework method, charting is the process of synthesizing the original data into the matrix. This process includes reviewing and summarizing the original data in forms such as original language as well as abstract terms (Ritchie & Lewis, 2003).

Value systems: A value system comprises sets of values through which individuals and organizations regulate their behaviour (Díaz et al., 2015). Such systems are influenced by factors including experiences, cultures, religion and political settings.

Instrumental value: Instrumental value refers to the direct and indirect contributions of nature's benefits to humans. These contributions are gained to achieve a good quality of life (Díaz et al., 2015). Examples can be the agricultural products for consumption and freshwater for drinking. In a specific value system—for example, the Total Economic Value framework—such values can be classified into non-use values too, including bequest value (preserving for future generation) and existence value (knowledge of continued existence) (Plottu & Plottu, 2007).

Intrinsic value: Intrinsic value refers to values that are inherent to nature, i.e., the fact that nature itself has values that are independent of human judgement. Such values are therefore beyond anthropocentric valuations (Díaz et al., 2015).

Relational value: Relational value refers to the type of value that exists in desirable relationships among people, and between people and nature. Its criteria might vary greatly across cultures and societies; therefore, it is a context-based state that consists of aspects such as access to food and water, good relationships between human and nature, and freedom (Díaz et al., 2015). Such values may imply trade-offs to obtain nature's benefits, therefore, they are not assessed in economic valuation frameworks.

Held value: Held value is interpreted as a series of desirable qualities or codes of conduct, such as fairness, courage, responsibility, equality which are broad and abstract (Chan, Gould, & Pascual, 2018).

Abstract

Values include a broad range of terms and are defined by different disciplines and across cultures and societies. In sustainable agricultural development, values are increasingly addressed as a means to understand the relationship between man and nature (Chan et al., 2018; Díaz et al., 2015). However, such debate is rarely discussed in the context of China. To narrow the lacuna, this paper takes Chinese ecological agriculture (CEA), a loosely defined concept, as an example to capture the role of values in sustainability transition, with a reflection on the relationship between man and nature. Specifically, this paper proposes three questions: What is the concept of ecological agriculture as expressed in current policies in China? What are the values in CEA? How do values affect the ecological agriculture implementation in China? To answer these three questions, this paper conducted a literature review and a case study on a local level. The results found that held values, instrumental values, intrinsic values, and relational values all played a role in CEA implementation in varying degrees. Other than that, policies and political rhetoric contributed to the CEA implementation and its social construct, respectively. The results also found that the concept of CEA evolves as the relevant policies develop. Currently, CEA combines traditional agricultural wisdom with modern innovative technologies to achieve economic, social, and ecological benefits while maintaining the carrying capacity of resources and contributing to increase in farmers' income. Recommendations were made to enrich the various aspects of CEA values, to refine the CEA concept, and to include various perspectives to facilitate the CEA development.

Keywords: Chinese ecological agriculture, values, nature, sustainability transition, policy

1. Introduction

1.1 Background

Values are complex constructs that can guide the thinking and action for a specific purpose. In sustainable development, values—sometimes referred to as a set of broad terms such as norms, regulations, etc.—play a key role because it reflects the underlying motivations and serves as an ethical foundation for actions. In the field of agriculture, values are classified as a set of ethical principles to understand the sustainability transformations in landscape stewardship (Bieling, Eser, & Plieninger, 2020). The meaning of different values is also important in understanding the relationships between man and nature because the way nature is treated reflects how man perceives themselves in relation to the environment. The power of different values varies in different contexts, and for different people and groups; therefore, when considering the roles of values in sustainability transition, it is necessary to identify what kind of values there are and to which extent they influence certain actions. Outside of values, there might be other factors that jointly influence the thinking and action for a specific purpose. For example, institutional arrangements, governance and anthropogenic assets are important elements that mediate the relationship between man and nature (Díaz et al., 2015).

Given that such debate is increasingly discussed in the western world, it is worthwhile to look into how it plays out in the context of China, especially when it is poorly discussed. To contribute to the study of how values, among other factors, play a role in the agricultural sustainability transition in China, this paper will take the Chinese ecological agriculture as an example for analysis.

In China, the number of environmental problems arising from intensive use of resources, especially in the agricultural field, has always been high since decades (Guo, 2015). One of the main drivers is population. Between 1954 and 1982, the population expanded almost two-hundred percent (National Bureau of Statistics of PRC, 2001a, 2001b), creating a huge demand on food production from limited arable lands. The traditional agriculture was unable to ensure food security for everyone, thus calling for a change into the conventional agriculture that utilises mechanization and external inputs to yield high production of food (Shi, 2004). The shift from traditional to conventional agriculture signified a dramatic change in the agricultural development at a huge expense of the environment, including vegetative destruction, soil erosion, water pollution and water resource depletion (Cheng, Han, & Donald C., 1992). In response to the dramatic environmental issues especially in rural areas, scholars and policy makers started to call for a transition from conventional agriculture to a more sustainable model that takes care of the environment during the 1980s (Ye, 1982). Emerging from the call is the proposal of Chinese ecological agriculture (CEA) that combines traditional agricultural practices with innovative and modern technologies, so as to ensure the plenty of food for fast growing population and alleviate the intensity of environmental pollution.

As a relatively new concept, CEA reflected an imperative demand for a greener environment, although relevant progress in the first two decades were rather limited in comparison to dominance of conventional agriculture. In the context of resource constraints, severe environmental pollution and the degradation of ecosystems, the notion of Ecological Civilisation Construction (ECC) has been brought up in the 18th National Congress of the Chinese Communist Party (NCCCP) in 2012 (Gu et al., 2013). ECC promotes respect, adaptation to and protection of nature, and integrates it into economic, political, cultural, and social developments, so as to realise the sustainable development of Chinese nation (Gu et al., 2013). The theme of ECC is to build a sustainable society that considers resource conservation, environmental protection, and establishes a harmonious relationship between man and nature. To this end, it included CEA among other sectors as elements necessary for sustainability transition. Meanwhile, the connotation of CEA has been refined by traditional agricultural wisdom as well as the recent concept of Two Mountains Theory (TMT) that deals with the relationship between man and nature (Xi, 2007).

1.2 Objectives and research questions

The analysis aims to find out how values influence the implementation of ecological agriculture in China, addressing the focus on the relationship between man and nature. To achieve this, three research objectives are proposed, each of which is followed by one research question.

The objectives are to:

(1) describe the concept of Chinese ecological agriculture (CEA);

(2) identify the values in CEA;

(3) discover how values affect the CEA implementation.

The research questions are as follows:

- What is the concept of ecological agriculture as expressed in current policies in China?

  Justification: Unlike the ecological agriculture that originates from the western world, CEA takes a different form and has its own connotations with Chinese characteristics. Therefore, it makes sense to understand what CEA stands for and how it is expressed in current policies in China. In doing so, it will also address the role of CEA that might contribute to the sustainability transition in China.

- What are the values in CEA?

  Justification: This question furthers the first question and examines what kind of values are included in CEA. From this point, it aims to find out different values that comprise the social construct of CEA. And since CEA relates to the interaction between man and nature, the way nature is treated reflects the type of values people hold towards it. Looking into this question will help understand the role of values in sustainability transition.

- How do values affect the ecological agriculture implementation in China?

  Justification: This question focuses on the role of values in sustainability transition from an empirical point of view. It will use a case study to examine how values as well as the political slogan "clear waters and green mountains are mountains of gold and silver," also known as the Two Mountains Theory (TMT), play out in the implementation of ecological agriculture. Focusing on one place where TMT applies, it also serves as an opportunity to identify the relationship between man and nature, thus contributes to sustainability transition and indicates further directions of its development.

The answers to the questions will verify whether the concepts about values are valid in the context of China, contribute to the international debate on sustainability transition, as well as point out new agendas for further research.

2. State of research

2.1 Concept of sustainability transition

According to the original definition of sustainable development, it is framed as the "development that meets the needs of the present without compromising the ability of future generations to meet their own needs" (Keeble, 1988). The four dimensions, i.e., society, environment, culture, and economy, are integrated into this concept. Sustainability is a paradigm towards an improved quality of life that balances the societal, environmental, cultural and economic considerations (UNESCO, 2019). Though the concept of sustainability transition might be subject to different interpretations, a common understanding of sustainability transition is radical and long-term processes that include multilevel changes in social-technical systems that shift towards a sustainable mode of production and consumption (Markard et al., 2012).

In agriculture, the sustainability transition therefore relies on efforts from all levels to contribute to a sustainable agricultural practice that ensures food security and mediates a harmonious relationship between human and the environment. The sustainability transition represents a shift in values, norms, and technology used in agricultural practices that are determined by the context of specific locations. In this sense, sustainability leads a general direction rather than a defined framework for agricultural practices and development (Shi, 2004). It is open for interpretations and lies in a dynamic state where values among other factors influence its development. Nevertheless, it is generally accepted that agricultural sustainability takes into consideration the carrying capacity of resources and environment, while ensuring food security, at least on a regional level.

2.2 Social values in sustainability transition

The role of values in sustainability transition, particularly regarding the man-nature relationship has been increasingly discussed. The IPBES framework is one example that deals with the connection of people and nature by identifying the role of values among other factors (Díaz et al., 2015). Intrinsic, relational, and instrumental values stressed in this framework derive from three key elements represented by nature; good quality of life; and nature's benefits to people, respectively.

Specifically, instrumental value refers to the direct and indirect contributions of nature's benefits to people. These contributions are gained to achieve a good quality of life (Díaz et al., 2015). Examples can be the agricultural products for consumption and freshwater for drinking. Utilising nature as an instrument to gain such benefits necessary for human survival and development therefore is a process characterised by instrumental value. In specific value system—for example, the Total Economic Value framework—such values can be classified into non-use values too, including bequest value (preserving for future generation) and existence value (knowledge of continued existence) (Plottu & Plottu, 2007).

Intrinsic value refers to values that are inherent to nature, i.e., the fact that nature itself has values that are independent of human recognition. Such values are therefore beyond anthropocentric valuations (Díaz et al., 2015). The intrinsic value also has a relation to ecological justice, which concerns the view that non-human beings have rights too (Baxter, 2004).

Relational value refers to the type of value that exists in desirable relationships among people, and between people and nature. Its criteria might vary greatly across cultures and societies, therefore, it is a context-based state that consists of aspects such as access to food and water, good relationships between man and nature, and freedom (Díaz et al., 2015). For example, the harmony between man and nature is pursued in Chinese indigenous religion Taoism (B. Zhao, 2009). Such values may imply trade-offs to obtain nature's benefits; therefore, they are not assessed in economic valuation frameworks.

The relational value is increasingly debated and refined from a conceptual level; particularly, it is distinguished from held value, which is also used in environmental discourses. The held value is interpreted as a series of desirable qualities or codes of conduct, such as fairness, courage, responsibility, and equality, which are broad and abstract (Chan et al., 2018). When a held value is applied in a specific context, it can be related to other values. For example, fairness is a moral quality but taking care of the nature because it's important for one's well-being connects a relational value; believing that nature has its own rights to live with man reflects intrinsic value; and using organic instead of chemical fertilizers to grow crops for sale relates to instrumental value.

To better understand these values and inform policy-makers, some scholars formulated a set of ethical principles to address them in the sustainability transformations in land stewardship (Bieling et al., 2020). The principles of Prudence, Justice, and Good Life refer to instrumental value, intrinsic value, and relational value, respectively. It needs to be pointed out, however, that such frameworks provide only a common ground to analyse the role of values in sustainability transition, because the complexity of interactions between human societies and natural world is not entirely recognized, particularly under different contexts.

2.3 The concept of CEA in relation to traditional, conventional and ecological agriculture

Historically, the agricultural change is broadly represented in three systems: traditional agriculture, conventional agriculture, and alternative agriculture. Each of these systems is associated with certain technologies, expectations, and interaction with the natural resources.

For many centuries, traditional agriculture has applied the basic techniques and methods of cultivation to provide sufficient food for people. Traditional agricultural practices are based on mostly smallholder systems that were capable of providing enough food for consumption. It is a sustainable system that is labour-intensive and aims for self-sufficiency. In China, the traditional agriculture practices reflected not only farmers' knowledge but also the wisdom of the indigenous religion *Taoism* since hundreds of years (Ge & Yuan, 2009). The Taoist view of nature is twofold: man and nature form an integral system in which man shall adapt to nature instead of disobeying the rules of it. The way to achieving this state is through the principle of Wu-wei, which means to adjust to nature and take appropriate actions based on the objective conditions and laws of nature. The Taoists believe that the relationship between man and nature shall be harmonious and mutually adaptive (B. Zhao, 2009).

Conventional agriculture took place since industrialization. It depends heavily on the mechanization and external inputs such as chemical fertilizers, creating much higher yield per unit use of land than traditional agriculture, and at a greater cost of environmental resources. It is an exploitative system that is chemical-based, market-oriented, and saves manpower. Green revolution is an example of how conventional agriculture took a toll on the nature. The overuse of chemical fertilizers drained the potential of soil fertility, resulting in a series of problems including biodiversity loss, pollution, etc., cutting off the connection between the community and the traditional agricultural practices, thus disrupting the harmony between man and nature (Shi, 2004). In China, the planned economy back then focused on unanimous and collective way of production and distribution, with the aim to maximize yield and feed the rapidly growing population (Shi, 2004).The socialistic political thinking prioritizes production and views agriculture as an industry that is to be instrumentalised for the benefit of human. This mentality, however, contradicts the Taoist-based wisdom on production which was reflected through Wu-wei.

As a result, a wave of sustainable alternatives emerged to combat this trend. Such practices aim to regenerate the ecosystems while ensuring food security and social justice for all. Ecological agriculture is one of them. The concept of ecological agriculture was defined by M. Kiley-Worthington in seven criteria: self-sustaining, diversified, small-scale, maximized yield, economically-viable, farm products sold directly to consumers, and acceptable from both aesthetical and ethical level (Kiley-Worthington, 1981). Certain techniques that are based on traditional agriculture are used to achieve this goal. Although ecological agriculture maximizes the yield with the techniques it is equipped, these small farming systems do not fit into the context of China. Therefore, when CEA was initially proposed, it has a different meaning than western ecological agriculture. CEA combines traditional agricultural practices with modern technologies that adapt to the local conditions.

2.4 CEA and its relevance to sustainability transition

The introduction of ecological agriculture is one example of sustainability transition that reflects a more harmonious relationship between man and nature. CEA is the Chinese edition of ecological agriculture which shares the common sustainability trait. Compared to conventional agriculture which relies on heavy external inputs for food production at a great cost of the environment, CEA follows the principles of ecology and combines the modern technology with traditional agricultural practices to form a sustainable agricultural system. It aims to improve people's quality of life, address the dilemma between human and the environment, and thus achieve economic, social, and ecological benefits (W. Wang, 1998).Therefore, promoting CEA is contributing to sustainability, and the shift from conventional agriculture to CEA is part of the sustainability transition process.

2.5 Ethical thinking in Chinese traditional agriculture

In China, virtues are an important aspect of human qualities. Since ancient times, Chinese start to recognize nature through the way they recognize each other, therefore, the ethical norms among people are extended to reflect that between man and nature. Nature is personalized and has feelings like human (Lai & Wang, 2019).

The Taoists are the advocates of ecological thinking. They proposed the equal treatment of the Universe (sometimes also referred to as Heaven), the Earth, and the people. The view of *Sancai*, which refers to these three elements, directly applies to the Chinese traditional intensive cultivation (G. Li, 2009). In this view, Heaven refers to the climate and Earth refers to the land; together they form nature, whereas man is the subject. Man is neither the slave nor the master of nature but a steward in the interaction with nature. Taoists believe that the relationship between man and nature shall be harmonious and work in concerted ways. One important aspect is to explore the laws of nature and adapt to it, so that man can also benefit (Ren & Lin, 2016). While man follows the laws of nature, he doesn't follow it passively. The intensive cultivation for example, requires the initiative of man that is based on the laws of nature and arranges agricultural activities according to the right time, in the right space and with the right use of resources (G. Li, 2009).

The Chinese ethical thinking of traditional agriculture shows in three principles: time suitability, land suitability, and carrying capacity of the land (Dong et al., 2018). These three principles are important for sustainable development in agricultural systems.

Time suitability: in agricultural production, it is important to wisely use the right time or season to grow crops and arrange the cultivation activities according to the natural laws of plant growth. To grow the right crops at the right time of the year is the natural way to adapt to nature (Dong et al., 2018).

Land suitability: this aspect follows the idea of adopting solutions according to the local conditions. A standard way of growing crops cannot be suitable to every land due to the variability in the soil conditions, etc. It depends on the local conditions to determine the varieties of crops for cultivation and the arrangement of agricultural activities. A reasonable method that suits the local conditions can ensure good use of resources and food production (Dong et al., 2018).

Carrying capacity of land: carrying capacity of land focuses on the potential production capacity of land while maintaining its stability. It is limited and requires conservation to keep and improve this capacity. This is particularly important during the fallow period when people stop growing crops and take care of the land to let it recover (Dong et al., 2018).

2.6 The Two Mountains Theory (TMT)

The Two Mountains Theory (TMT) is a concept as well as a political slogan proposed by President Xi in 2005. It states that "clear waters and green mountains are as valuable as mountains of gold and silver" (Xi, 2007). It is a metaphor for a win-win state of economic development and environmental protection. Clear waters and green mountains refer to the natural environment, whereas mountains of gold and silver refer to the economic growth derived from environmental resources.

President Xi elaborated the concept of TMT in 2006: "The harmony between man and nature, and the harmony between economy and society, are both pursued by mankind. Put in common terms, we want 'two mountains': the mountains of gold and silver; and the clear waters and green mountains" (Xi, 2007). These two mountains can be contradictory, as well as dialectical to each other. In practice, three stages are necessary for understanding the interaction of the two mountains: in the first stage, man exploits nature and derive benefits without considering the availability of resources and carrying capacity of the environment; the clear waters and green mountains are sacrificed for mountains of gold and silver. In the second stage, as the environment deteriorates from exploitation, resources become scarce, and economy growth cannot be sustained the way it was before, so man starts to realize the importance of conserving nature and the fact that they rely on nature; they want to protect the clear waters and green mountains as well as keeping mountains of gold and silver. In the third stage, man realizes that to keep the two mountains, it is necessary to change development to a sustainable mode, i.e., moving towards sustainability transition. When economic development is derived sustainably from natural resources, the clear waters and green mountains can continuously provide mountains of gold and silver—and they are the mountains of gold and silver (Xi, 2007).

TMT is based on the Marxist view of nature (harmonious coexistence between man and nature) and draws on the Chinese traditional wisdom (S.

Gao, 2019). It emphasizes the harmonious interaction between man and nature during economic development, therefore achieving a win-win situation. In this concept, man and nature interact with each other and stay in a dynamic state. TMT is included in the 19th Communist Party of China National Congress report in 2017, and comprises one of the important principles of promoting ECC in 2018 (ChinaDaily, 2017).

Though clear waters and green mountains are as valuable as mountains of gold and silver, it is important to point out that the former can bring the latter, but the latter cannot buy the former (Xi, 2007). When there is conflict between environmental protection and economic growth, the long-term ecological benefits are more important than short-term economic growth, i.e., keep clear waters and green mountains instead of changing them. The fine line to consider is what nature provides and how to adapt it to local conditions.

3. Methodology

This paper applies a qualitative and inductive method and examines the research questions through a literature review and a case study.

3.1 Literature review

For the literature review, this paper adopts traditional literature review, which aims at providing a critical written account of the current state of research on a selected topic. Since the role of values in sustainability transition in China is still an emerging topic that lacks attention, it is meaningful to adopt this approach. In this paper, the literature review will be processed in five steps.

- First, the scope of literature review is defined, for this part, it is closely related to the research questions.
- Second, the selection of databases is identified.
- Third, the initial search is conducted for sources in databases that are possibly relevant to the topic.
- Fourth, the final sources for literature review are identified by reading and analysing the title, abstract, and keywords of each paper. For this part, a set of exclusion criteria are applied.
- Fifth, the result of the finalised sources is presented, analysed, and interpreted. Particularly, the results of the data will be discussed in relation to the research questions.

3.1.1 For the scope of literature review

The scope of literature review focuses on values and ecological agriculture in China. Specifically, the scope revolves around the three research questions: What is the concept of ecological agriculture as expressed in current policies in China? What are the values in CEA? And, how do values affect the ecological agriculture implementation in China?

3.1.2 Selection of databases

This paper chooses two scientific databases, i.e., Scopus and CNKI (Chinese equivalent of Scopus) for research papers mainly, and other publications, such as thesis. The language of the literature will thus be in English and Chinese. The reason for including Chinese literature is because of the scarcity of English literature available for looking into this topic and the fact that Chinese literature can be of great help to contribute to the international debate in this regard.

3.1.3 Conduct of initial search

For this part, keywords are used for identifying possible literatures in the databases. The selected topic-related keywords include: value, social value, nature, sustainability, ethics, CEA, ecological agriculture, China, sustainable development, wisdom, religion, TMT, policy, case study, and implementation. Different combinations of keywords in both languages will be used in order to locate relevant papers as many as possible. This will create a list of identified papers for further selection. Specifically, for each of the research questions, different keywords, and combinations of keywords are used for identifying the papers. It is important to note that the same papers might appear in these searches because the difference among the combinations of keywords are moderate. The search results will be presented in tables.

Table 1 lists keywords used in the search for papers pertaining to each of the three research questions. Additionally—and importantly—different combinations of keywords are presented.

Table 1. Keywords for search in databases

Research questions	Keywords	Combination of keywords for search
What is the concept of ecological agriculture as expressed in current policies in China?	China; ecological agriculture; policy; sustainable development; sustainability	China, ecological agriculture, policy; China, ecological agriculture, policy, sustainable development; China, ecological agriculture, policy, sustainability
What are the values in CEA?	China; ecological agriculture; value; social value; ethics; wisdom; religion; nature	China, ecological agriculture, value; China, ecological agriculture, ethics; China, ecological agriculture, wisdom; China, ecological agriculture, social value; China, ecological agriculture, religion; China, ecological agriculture, nature
How do values affect the ecological agriculture implementation in China?	China; ecological agriculture; value; TMT; implementation; case study	China, ecological agriculture, value, implementation; China, ecological agriculture, value, case study; China, ecological agriculture, TMT, implementation; China, ecological agriculture, TMT, case study

As presented in Table 1, for each research question, 3-6 sets of keywords were used for searching relevant papers. The keywords are selected based on its relevance to the topic in order to locate as many papers as possible. The keywords are searched set by set, i.e., they are not used altogether for one search.

3.1.4 The analysis of the literature

Selection of literature:

From the search results, an analysis of the article abstract/keywords/title is carried out to finalise the topic-relevant articles. The exclusion criteria include relevance to the research topic and geographical location of study. Given the limited sources of this topic, date range is not considered in the exclusion criteria.

After the initial search from Scopus, a total of 61 papers were found. The number of papers from each search is listed in Table 2.

Table 2. Initial search results from Scopus based on combination of keywords

Research question	**Keywords**	**Search results (the number of articles)**
What is the concept of ecological agriculture as expressed in current policies in China?	China, ecological agriculture, policy	29
	China, ecological agriculture, policy, sustainable development	9
	China, ecological agriculture, policy, sustainability	4
What are the values in CEA?	China, ecological agriculture, value	39
	China, ecological agriculture, ethics	1
	China, ecological agriculture, wisdom	1
	China, ecological agriculture, social value	0
	China, ecological agriculture, religion	0
	China, ecological agriculture, nature	5
How do values affect the ecological agriculture implementation in China?	China, ecological agriculture, value, implementation	4
	China, ecological agriculture, value, case study	6
	China, ecological agriculture, TMT/Two Mountains Theory, implementation	0
	China, ecological agriculture, Two Mountains Theory, case study	0

Table 2 shows the initial search results from Scopus database. The majority of papers were located based on keywords search that included "policy" and "value." However, the word "value" can refer to many different values, such as economic value, social value, ecological value, etc.; therefore, it is important to apply exclusion criteria to select those pertaining to the topic of this paper. After the screening of the 61 papers based on the exclusion criteria, a total of 22 papers were identified for literature review. The exclusion process typically removed papers that only use economic models and/or are irrelevant to the research questions. Among the 22 papers, only 9 papers were accessible from the university library and the rest of them were either not accessible or were not found, according to the Scopus search. Nevertheless, out of the 13 papers that were not accessible from Scopus, 8 papers were accessed from the CNKI portal, making the total number of papers 17 for literature review (11 English papers and 6 Chinese papers).

After the initial search from CNKI, a total of 138 papers were found. The number of papers from each search is listed in Table 3.

Table 3. Initial search results from CNKI based on combination of keywords

Research question	Keywords	Search results (the number of articles)
What is the concept of ecological agriculture as expressed in current policies in China?	China, ecological agriculture, policy	ecological agriculture (生态农业, N=40); ecological agriculture development (生态农业发展, N=20); sustainable development of agriculture (农业可持续发展, N=13); ecological agriculture construction (生态农业建设, N=9); Chinese ecological agriculture (中国生态农业, N=8); agricultural policy (农业政策, N=7); ecological civilisation (生态文明, N=6); modern ecological agriculture (现代生态农业, N=5); policy recommendations (政策建议, N=5); efficient ecological agriculture (高效生态农业, N=3); ecological agriculture mode (生态农业模式, N=5)
	China, ecological agriculture, policy, sustainable development	sustainable development of agriculture (农业可持续发展, N=13); ecological agriculture (生态农业, N=13); ecological agriculture development (生态农业发展, N=7); ecological civilisation construction (生态文明建设, N=6); ecological agriculture mode (生态农业模式, N=3); modern ecological agriculture (现代生态农业, N=2); study review (研究述评, N=2); ecological agriculture construction (生态农业建设, N=2)
	China, ecological agriculture, policy, sustainability	sustainable development of agriculture (农业可持续发展, N=4); ecological agriculture (生态农业, N=4); ecological agriculture mode (生态农业模式, N=3); ecological agriculture development (生态农业发展, N=3); sustainability (可持续性, N=1); ecological agriculture construction (生态农业建设, N=1); Chinese ecological agriculture (中国生态农业, N=1); production mode of ecological agriculture (生态农业生产模式, N=1)

Table 3 cont. Initial search results from CNKI based on combination of keywords

What are the values in CEA?	China, ecological agriculture, value	ecological agriculture (生态农业, N=10); modern ecological agriculture (现代生态农业, N=3); ecological agriculture construction (生态农业建设, N=3)
	China, ecological agriculture, ethics	ecological agriculture (生态农业, N=4); ecological agriculture development (生态农业发展, N=2); ethical appeal (伦理诉求, N=1)
	China, ecological agriculture, wisdom	ecological wisdom (生态智慧, N=5)
	China, ecological agriculture, social value	agricultural transition (农业转型, N=5)
	China, ecological agriculture, religion	0
	China, ecological agriculture, nature	ecological agriculture (生态农业, N=37); ecological agriculture mode (生态农业模式, N=7); ecological agriculture development (生态农业发展, N=7); modern ecological agriculture (现代生态农业, N=6)
Research question	**Keywords**	**Search results (the number of articles)**
How do values affect the ecological agriculture implementation in China?	China, ecological agriculture, value, implementation	efficient ecological agriculture (高效生态农业, N=1);
	China, ecological agriculture, value, case study	Poverty reduction by industry (产业扶贫,N=1)

Table 3 cont. Initial search results from CNKI based on combination of keywords

	China, ecological agriculture, TMT/Two Mountains Theory, implement-ation	0
	China, ecological agriculture, Two Mountains Theory, case study	0

Table 3 illustrates the initial search results from CNKI. The results are classified into different subtopics with the number of papers and a Chinese translation of the subtopics included in brackets. The reason for presenting the results this way is because the searching function varies from and is not as straightforward as the one from Scopus, i.e., the searching results from CNKI based on keywords search are fuzzier and are classified into different subtopics. In a way, it is good for looking into a poorly defined topic because more relevant papers can be identified. However, some of these subtopics can be unhelpful in producing the desired information, therefore, a further selection of relevant subtopics is conducted to narrow down the scope of the initial search results. The chosen subtopics are shown in table 3.

After selecting the papers based on the exclusion criteria, a total of 38 papers were identified for review. However, due to the same reason, only 25 papers were accessed and will be used for literature review.

After further analysing the contents of the papers, two more were removed because they were irrelevant to the research questions. They were not removed earlier because the titles were confusing. This makes a total of 23 papers for literature review from the finalised search. As the literature review analysis continues, additional literature is included for support of important findings. These literatures include governmental reports and conference proceedings.

3.2 Case study

3.2.1 Selection of the case

While the literature review illustrates the various concepts used in this paper, such as values, sustainability transitions, CEA, etc., which deals with the topic from a conceptual level, the case study will provide an empirical look at how the values affect the CEA in practice and identify if there are any other influencing factors. Specifically, it seeks to identify how the TMT affects the implementation of CEA. To this end, it will locate one of the regions where TMT applies. Given the timeframe and the resources available, the author located one place in Sichuan province, southwest China.

The case study focuses on the rice-fish-crab farming system in Youjun village in the city of Xichang, southwest China. One farming cooperative from this village initiated the first non-chemical farming in China in 2016, which focused on the use of bio-organic fertilizer to engage in ecological agriculture and has become well-known in the last two years. The cooperative introduced several modes of ecological agriculture including the rice-fish-crab system and rice-duck system. Given the availability of time and resources, the case study decides to focus on one system, i.e., the rice-fish-crab farming system.

The geographical location of the village is shown from Figure 1 and Figure 2. Fig. 1 gives an orientation from a country-wide view, and Fig. 2 shows a closer look at the village from a satellite map view.

Figure 1 shows the location where the case study takes place. The city of Xichang is the capital city of the Liangshan Yi Autonomous Prefecture, Sichuan province.

Figure 2 shows a satellite view of Youjun village, which is located at the southwest part of Xichang city. Agriculture comprises the majority part of the village economy. There are 1,533 hectares of arable land, with a ratio of 0.1 hectare per capita (Huang, 2016). The main crops for cultivation are rice and wheat. The red dot and its vicinity area are the farm field that belong to this cooperative.

Rice-fish-crab farming system

The rice-fish-crab farming system is based on the rice-fish farming system, which draws on the wisdom of traditional agriculture and has existed for

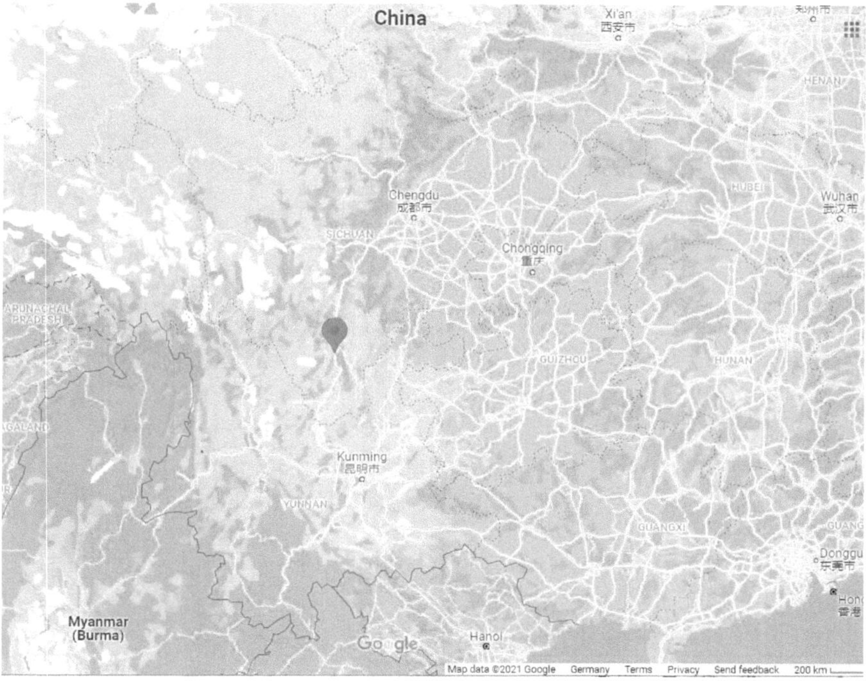

Figure 1. The geographical location of the city of Xichang (Source: Google Maps)

Figure 2. Satellite map of Youjun village (Source: Google Maps)

hundreds of years (Lu & Li, 2006). Figure 3 briefly explains the working mechanism of such farming system.

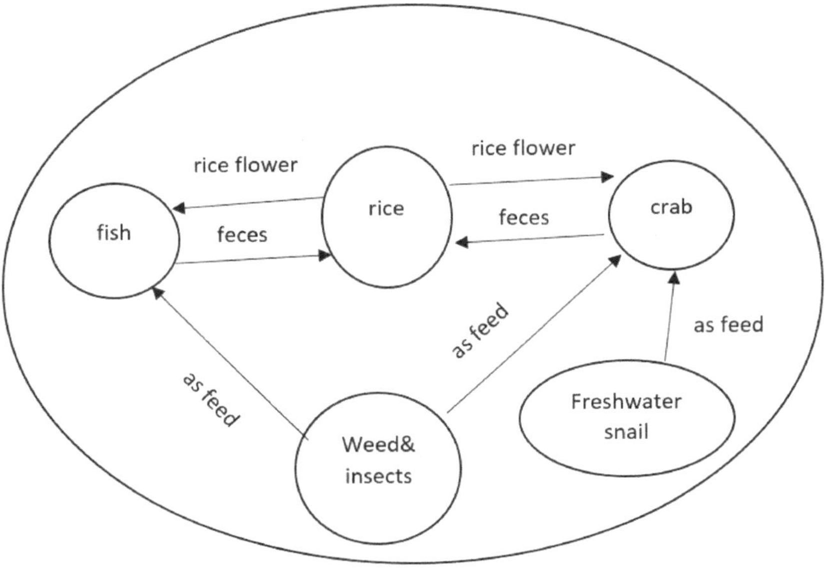

***Figure* 3.** The working mechanism of rice-fish-crab farming system

Figure 3 illustrates the relationships and nutrient flow in rice-fish-crab farming system for the farm under study. Both the fish and crabs feed on the weeds and insects in the water, and crabs also feed on the freshwater snails. The faeces produced from fish and crabs serve as nutrients necessary for the rice growth. Fish and crabs contribute to the increase in dissolved oxygen as they move around. During the flowering time, the crabs and fish feed on the rice flowers. In this way, the resources in the system are efficiently used and the use of pesticides is significantly decreased.

To study how values play a role in this CEA model, an in-depth interview was carried out with one farmer representative, who is also the initiator of the rice-fish-crab farming in Youjun village.

3.2.2 Selection of the interviewee

Sampling rule for the in-depth interview: given the research questions, it is important to select those who can provide the maximal information needed for addressing such questions. This paper uses purposive sampling, focusing on those who are able to answer the questions in a relatively complete and accurate way.

The selection of the interviewee is based on two aspects. First, the interviewee shall be an experienced farmer who is managing ecological agriculture and knows not only the farming techniques but also relevant agricultural policies, i.e., a skilled farmer that understands policies. Second, the interviewee is fully aware of the purpose of the interview and is willing to participate. For a successful implementation of an in-depth interview, three ethical rules must be followed: informed consent, minimal harm, and anonymity.

The process of identifying the right interviewee took three weeks. This is due to several reasons:

- First, during the COVID-19 pandemic, the interview is designed to be conducted online.
- Second, the indirect contact with farmers, such as searching for contact information online through local newspapers and media, was challenging because the response rate was rather low.
- Third, even if the right interviewee were found, persuading the person to participate in this interview was another challenge.

With such uncertainty and complexity in hand, the interviewer resorted to resources and connections he had in China. With an actual farm visit paid by two local contacts in China, the interviewer was able to connect with the identified farmer representative who later agreed to participate in the interview. One key point that helped facilitate this agreement was the fact that the interviewee and interviewer both come from the same region and speak Sichuanese Mandarin, a local dialect in Southwest China (this was also proved later by the farmer during the interview).

3.2.3 Interview process

With the informed consent from the interviewee, the interview was conducted online and the audio recording was documented, obeying all the formalities for carrying out the in-depth interview, including the rules of minimal harm and anonymity. The language of the interview was Sichuanese Mandarin, a dialect spoken in Southwest China.

As agreed by both parties, the interview began on 13th of July 2021 at 5:00 a.m. and ended at 6:30 a.m. (CET time). At the beginning of the interview, the interviewer introduced himself and welcomed the interviewee with a brief description of the theme and formalities. During the interview, seven interruptions occurred that inevitably compromised the quality of some important discussions. Because of the same reason, the recordings were made twice, totalling 71 minutes. At the end of the interview, the interviewer thanked the interviewee again and they agreed to keep in contact when necessary.

The interview questions are shown in Table 4.

Table 4. The interview questions

Question number	Text
1	When was the cooperative established and how many farmers have joined it?
2	Did you organize and set up the cooperative? Or how have you made them join it?
3	Did you only grow rice? Was it monoculture?
4	Are the baby fishes also purchased from Zhejiang province, or you use the local baby fishes?
5	Do you grow rice once a year or several times a year?
6	I wonder if you can describe the situation of harvesting the fish and crabs. Is this activity only involving farmers, or are citizens also invited to participate?
7	How were you able to gather the farmers together to engage in this type of ecological agriculture? Did they look for you or you persuaded them?
8	For the use of such technology, I wonder if the national or local government provided any subsidies or policy incentives?
9	Since the rice is grown once a year, how do you protect the land when it is in the fallow period?
10	How did you come to know the non-chemical farming practice and the rice-fish-crab system? Was it out of interest and you looked for it somewhere?
11	I wonder if you ever thought about subsidy before you started rice-fish farming system, or you came to know the subsidy after you started it and you felt more motivated to continue doing it?
12	I was asking if you received the subsidy after you started the non-chemical farming system, or you knew the subsidy first and then decided to do this?
13	So it means you applied for it later, or they issued you the subsidy when they came to know you were doing such farming?

Table 4 cont. The interview questions

Question number	Text
14	Do you know if there are any specific policy documents that support the non-chemical rice farming? Are there more local policies or national ones?
15	For this policy document, do you have it or do you know where one can find it? Perhaps from online?
16	Ok, and it is a national rather than local policy, right?
17	How did you persuade the other farmers to join the cooperative? Did they volunteer to participate when they heard this project? Or after your persuasion…
18	Yes, yeah, did the other farmers grow only rice in their own field? What did they grow?
19	So it means that the other farmers contribute part of their own land to your cooperative for implementing such farming practice?
20	How did they feel after they joined you? I mean, except for the economic profit, did they feel that engaging in non-chemical rice agriculture meant something more than just income growth? Perhaps they also feel attached to the land? Or they simply feel the main thing is to earn more money?
21	Do you remember what kind of environmental problems were there before the ecosystems were recovered? Any specific aspect?
22	Was there no air pollution etc.?
23	In your daily life, except for work where you go with other farmers to the field, are there any other moments when you feel like going there for a walk, because of the beautiful view etc.?
24	Do you go for a walk (in the field) to relax yourself?
25	Do you have further plans regarding the non-chemical farming? Have you thought about expanding it?
26	What kind of techniques have you used for conservation of the field?
27	Are there also farmers who are willing to participate in this in-depth interview with open questions?

The questions in Table 4 were asked during the interview. These questions revolve around the research questions and try to find out if there are other influencing factors in the implementation of CEA. Except for a few pre-defined questions that were used to get general information about the farm, most of the questions were asked depending on how the interviewee responded the previous questions and were formulated as plain as possible.

Some questions were repeated a couple of times because of the poor internet connection and/or interruption on the interviewee's side.

3.2.4 The analysis of the interview

For the analysis of the case study, the method of Framework is applied. It is a qualitative analysis method developed by two researchers, Jane Ritchie and Liz Spencer, at the National Centre for Social Research in the UK in late 1980s (Ritchie & Lewis, 2003). It was originally developed for large-scale social policy research by qualitative researchers. The name Framework comes from "thematic framework," which is used to organize data and cluster them according to codes and categories. In this way, a framework comprises a series of categories that are divided by codes. Codes and categories will be refined through the process of familiarisation with the data. When it is considered comprehensive, each of the categories is displayed in a matrix where each respondent is allocated a row and each column is represented by a code (Ritchie & Lewis, 2003). The summarized data are put into the cells. To achieve this, the Framework method consists of 6 parts: transcription, familiarization with the interview, coding, developing a working analytical framework, charting data into the framework matrix, and interpreting the data. Given that there is only one participant for this interview, there is no comparison to be made among multiple participants; therefore, the results are simply presented in tables for a clear understanding.

Part 1: Transcription

The two recordings were used for verbatim transcription. During the transcription stage, some parts of the recordings were played multiple times to ensure the accuracy of the transcript. When it was complete, the transcription was sent to the interviewee for review. This review process was necessary because multiple technical/buzz words were used by the interviewee and there were several moments when the interviewer and the interviewee spoke simultaneously. After the review process, a translation process from Chinese to English was conducted. The whole process serves as a good opportunity to familiarize with the data.

Part 2: Familiarization with the interview

Using the finalised transcript to become familiar with the interview is important for ensuring the overall quality of the analysis. To achieve this, researchers in the same project shall all familiarize themselves with the interview. For this study, since the transcriber, translator and analyst are the same person, this process is done in combination with the transcription part.

Part 3: Coding

After familiarization, the finalised transcript was imported to QDA Miner, a computer-assisted qualitative data analysis software, for coding. The coding process applies open coding (Corbin & Strauss, 1990), which is quite useful for inductive approaches that look into something not yet well understood. This is also helpful for capturing all possible data relevant for the research questions.

For the open coding: the transcript is carefully read line by line, divided into meaningful parts, usually in the form of sentences, and labelled with a code that summarizes the meanings. After coding all the data, codes are clustered into categories according to their common/similar traits. A code can be a single word or set of words with annotations attached to it, i.e., the codes are condensed from the transcribed interview and are described in phrases and/or short sentences. Anything that might be relevant from different perspectives are coded, too. For this reason, the number of questions and times of interview interruption are also documented because they will be later used for further reflections, such as understanding farmer's language.

The initial results from the open coding captured 77 inductive codes. Some of these codes are exactly parts of the texts; others are refined and summaried in a more abstract way. The codes generated from the open coding are correlated by analysing the similar and/or common traits, i.e., common themes that several codes share. The QDA Miner is of great help in this stage as it allows quick retrieval of the indexed data for each code. Based on these traits, 77 codes are clustered into 14 categories, i.e., codes that are conceptually related are within the same category, and two separate codes that do not fall into any of these categories.

Since this paper captures only one interview, the code formulation process is done by the author without any peer reviews. This is also due to time constraints. Inevitably, the codes are subjective, but they are based on reasonable and inductive thinking thus not biased.

Part 4: Developing a working analytical framework

All the codes and categories form the analytical framework. The framework includes each of the codes explained in a breif description, accompanied by the number of times and the percentage out of all codes. The categories are highlighed in bold letters to distingush from codes. A total of 14 categories and 2 separate codes is presented in the framework in the appendix (see Table 24).

4. Results

4.1 Results from the literature review

The results of the literature review are divided in three parts, i.e., the theoretical development of CEA, the ethical thinking in CEA, and its policy development.

4.1.1 Theoretical development of CEA

There are multiple descriptions regarding the definition of CEA. The mainstream one is proposed by Chinese ecologist Qianji Ye in 1981 and developed thereafter by different scholars. Ye started from the view that in agricultural systems, man, biological organisms and the environment are in a contradictory unity and that man takes a dominant role. Man needs a good living condition, but when it comes to production, man ought to take the initiative to change the environment in a way that promises production while obeying the laws of nature. It is in this sense that the current and future generations can enjoy a balanced and efficient ecosystem, i.e., the Chinese ecological agriculture (Ye, 1982). Initially, CEA was largely constructed by theories of ecological economics with the idea that economic development and environmental protection could go hand in hand (Shi, 2003). At that time, China was economically poor and there was a large population relying on the limited land use. The CEA therefore draws on various forms of the western ecological agriculture and adapts the theory and methods to its own context. The precondition for CEA is that it must fit into the actual conditions in China, being that the arable land in proportion to the overall land size is small; that there are large areas of mountains with few forests; that there is a large population but the rural labour is not used efficiently; and that the water and soil erosion is severe across the overall landscape (Ye, 1982). Therefore, the forms of CEA shall target and adapt to these conditions. To achieve this, Ye argues that CEA needs to combine traditional agriculture with modern technologies, which makes full use of energy flow and regenerate the substance circulation, so as to achieve an integral agricultural system that takes care of ecology and economic development, as well as the economic, social and ecological benefits (Ye, 1982).

Shijun Ma argues that CEA follows not only the laws of nature but also the laws of economy and aims for high production of food and increase in income for farmers. Based on this, Ma introduced the concept of ecological engineering to CEA, arguing the combination of system engineering methods can help achieve the economic, social, and ecological benefits (Ma & Li, 1987).

The definition of economic, social and ecological benefits of CEA is yet to be clearly defined. However, in general, these three benefits can be explained in the following:

- Economic benefits: CEA combines modern technologies with traditional agricultural practices in the production and processing of food. Support from the policy level helps marketing of food; meanwhile, the farmers earn more income (M. Zhao, 2019).
- Social benefits: in general, social benefits of agriculture include social stability, food security, contribution to local employment, cultural values, and expression of local customs (Luo, 2018).
- Ecological benefits: for ecological protection, CEA aims to follow the laws of nature, improve the capacity of ecosystem services, and decrease the negative impacts on it, while maintaining the balance in the system (M. Zhao, 2019).

After Ye and Ma, several more scholars contributed to the concept of CEA. These contributions reflect multi-faceted nature of CEA, including: the joint development of economy and environment protection; the low-input and high-output feature; the integration of rural development, environment conservation and efficient use of resources (Shi, 2003; M. Zhao, 2019). These aspects can be further explained in five points as follows:

- First, CEA pursues the integration of economic, social, and ecological benefits. It focuses on maximizing production by efficiently using resources while protecting the environment.
- Second, CEA combines the human interference with natural regulation of agricultural ecosystems. Unlike the developed countries, China has a large population, experiences increasing loss of water and soil resources, and faces deteriorating ecological environment in many regions. Therefore, the development of CEA aims to address these issues and meet the regional development goals which include improving famers' income (Zhu, 2009). In doing so, CEA focuses on the natural regulation of agricultural ecosystems, complemented by human interference that follows the laws of nature and organizes the production for the long-term benefit of people and nature by using multi-disciplinary methods

drawing from ecology, economics, environmental science, and systems engineering (Ma & Li, 1987; Zhu, 2009).

- Third, CEA has distinct regional features. The territory of China spans a wide range of distance, and the natural conditions, resource endowment and the level of economic and social development vary greatly across regions. Therefore, the development of CEA centres on the adaptation to local conditions with appropriate use of local resources (Zhu, 2009).
- Forth, CEA combines modern technology with traditional agricultural wisdom. The development of CEA is heavily influenced by traditional agriculture. A lot of practices related to intensive farming without mechanisation, as well as environmental protection, are absorbed in CEA. With the help of modern technology, CEA is an efficient system that integrates economic, social, and ecological benefits (Ma & Li, 1987; Zhu, 2009).
- Fifth, CEA emphasizes the multi-functionality of agriculture. To fully explore the potential of agriculture under a sustainable mode, as well as to meet the market needs and regional development goals, CEA also extends to non-agricultural rural industries such as new energy, fertilizers, and food processing etc (Cheng et al., 1992; Zhu, 2009).

Thus, the concept of CEA is further improved and its meaning is updated, i.e., it is based on maintaining and improving the production capacity of ecosystem, using modern technologies and management while aiming for rural development and increase in food production and farmer's income. In this way, the economic, social, and ecological benefits are achieved (Liu et al., 2017).

To sum up the important aspects of CEA, some authors formulated four rules for a successful CEA mode. They are condensed into four adjectives: circular, low-carbon, green, and efficient (Y. Zhang, Lin, & Li, 2015).

- Circular: CEA makes full use of the energy and circulate the substances and energy flow within the system. An example is the rice-fish farming system.
- Green: CEA relies on symbiosis relationship among different components of the system to improve the efficiency of resource use and the quality of crops. It also contributes to pest control

and reduces the use of chemical fertilizers and pesticides in the production process.

- Low-carbon: CEA consumes less energy and contributes to less pollution and carbon emission than conventional agriculture.
- Efficient: CEA aims for high efficiency of input and output, as well as high efficiency in use of resources.

Apart from the mainstream definition of CEA, some scholars hold a different view. They believe that CEA is an artificial system regulated within natural environment, and that all substance flows in CEA should be in a closed circle and rely on only a few external input to produce maximized output (Xiao, Zhou, & Sun, 2005). Moreover, some scholars argue that CEA is a type of agricultural mimics that simulates natural ecosystems (Lin & Chen, 2019). In this view, it is a sustainable agricultural system that utilises ecology and relevant principles to design and manage agriculture. It is also a new way of connecting agriculture with social groups that support sustainable agriculture and is a bottom-up approach that contributes to social movement (Shi, 2003). However, it remains unclear what kind of principles and structure of design are there. The potential of CEA as a social movement to be recognised by the government and the society has not been elaborated in detail, rather in a more general way.

Though CEA is described in various ways, the meaning of CEA and its implications are loosely delimited (J. Zhang & Luo, 2005). Some scholars argue that CEA is simply farming without pesticides and replaces chemical fertilizers with organic ones (Shao, Chen, Xiao, Zhu, & Li, 2019), while others believe that since the definition of CEA is too broad, and a lot of agricultural practices are related to it, they all can be considered as ecological agriculture (Qi, Si, & Scott, 2021; Tan, 2019).

4.1.2 Ethical thinking in CEA

The ecological thinking derived from traditional agriculture has implications for and is applicable to CEA. The sustainable development of agriculture shall not only depend on the use of modern technologies but also combine the traditional agricultural wisdom and practices to ensure the goal of stable and safe production of food. The three principles of time suitability, land suitability, and carrying capacity of land reflect the idea of man staying in harmony with nature (Dong et al., 2018). For agriculture systems, the harmony between man and nature means to balance the relationship between different elements of the system while bearing in mind that man and nature are not in opposition to each other (G. Li, 2009). This paves the way for achieving economic, social, and ecological benefits of CEA (Dong et al., 2018).

The ecological view of CEA is that man comprises part of the nature and is the subject of labour as well as the proactive regulator of the agricultural system. Man takes proactive measures to maintain the balance and stability of this system, which is emphasized in the harmony between man and nature. Since man is part of the nature and the regulator of the system, on one hand he learns from experience that he shall respect nature, on the other hand he takes the initiative to engage in agricultural production in a way that follows the laws of nature (Hao & Zhang, 2014). Under the natural conditions, man learns that growing crops at the right time of the season will guarantee harvest, while growing them at the time that does not enable the growth will lead to decrease in harvest or even no harvest. This also implies that it is important to leave enough time for the conservation of land, which has practical meanings for construction of CEA, that man shall work within the carrying capacity of land for long-term benefits (Hao & Zhang, 2014). Tan echoes with the idea of harmony between man and nature, arguing that only with this as its ecological wisdom, CEA is able to adapt to the social and economic changes and absorb the modern technology for sustainable development in rural areas (Tan, 2019).

As Ye initially proposed the concept of CEA, he argued that man takes the dominant role in the agricultural system. In order to manage this system and produce more food and materials for the increasing needs of the population, man ought to wisely use modern technologies and apply the rule of adaptation to the local conditions to change nature and create a reasonable ecosystem that suits human needs. He also pointed out that it is necessary to follow the laws of nature and the economy (Ye, 1982).

As Ye refined this concept in 1987, he mentioned that for a strategic transformation of CEA, an important aspect is to reconsider the relationship between man and nature, that CEA shall recognize that man is part of the nature instead of its ruler (Ye & Luo, 1987). Therefore, instead of dominating nature, man ought to be an attentive steward and form a harmonious relationship with nature because it is a living system that differentiates from mechanical stuff created by man, i.e., nature has its own life that shall be treated with care (Ye & Luo, 1987).

4.1.3 Traces of CEA expressed in policies

A compilation of CEA-relevant policies is presented below in chronological order:

From 1973 to 1981: indirect CEA policies appear at this stage where the ecological problems in agriculture and in rural areas started to get partial attention. Although the concept of CEA was not proposed during this stage, some policies were passed to protect rural ecology and ecosystems, therefore, they partially paved the way for establishing CEA. Such policies deal with three issues mainly, i.e., treatment of water pollution and drinking water in rural areas, environmental resource protection, governance of pollution from use of pesticides (M. Zhao, 2019).

Since CEA was firstly introduced in 1981, the No.1 Central Document, a policy document issued annually by the central government of China, started to emphasize the transition into the type of agriculture that makes full use of the advantages of traditional technologies, and resorts to modern technologies in a way that saves resources while promising production and protecting the environment (W. Li, Liu, & Min, 2010). A series of reports and governmental meetings also mentioned from a strategic point of view that China shall promote CEA and actively protect the environmental resources necessary for developing it.

From 1982 to 1992: ecological policies supporting CEA came into being. These policies target environmental protection and an overall governance of ecological problems in rural areas. Scholars such as Qianji Ye started to define CEA and this helped lay down the theoretical foundation of relevant policymaking. During this period, policies focused on two aspects. The first aspect is about the governance of agricultural ecosystems. Multiple CEA-related policies were made, including The Water Pollution Prevention and Control Law of the People's Republic of China (1984), Recommendation on Development of Ecological Agriculture and Reinforcement of Agricultural Environment Protection (1985), and Water and Soil Conservation Law of the People's Republic of China(1991) (M. Zhao, 2019).

Particularly, the Recommendation on Development of Ecological Agriculture and Reinforcement of Agricultural Environment Protection issued by the State Council clearly stated that developing ecological agriculture is an important approach to realizing the two transitions, one being transitioning from planned economy to market economy, the other being transitioning from rural to urban society (M. Zhao, 2019). The second aspect is about strengthening policies that target environmental protection in rural areas. From 1982 to 1986, the State Council issued five No.1 Central Documents that provided policy support for CEA development, see Table 5.

Table 5. Policy support for CEA development

Year	Title of No.1 Central Document	Relevant Content
1982	Minutes of Meeting on National Rural Work (全国农村工作会议纪要)	Agriculture activities shall aim to use low-investment and low-energy cost to generate high benefits and protect the environment. It is important to collect resource surveys and divide agricultural regions for tapping into resources in an appropriate way (State Council of PRC, 1982).
1983	The Several Issues Concerning Current Rural Economic Policies (当前农村经济政策的若干问题)	Use natural resources appropriately, maintain good environment, motivate the public to engage in ecology protection (Central Committee of the Chinese Communist Party, 1983).
1984	The Notice on 1984 Rural Work (关于一九八四年农村工作的通知)	Develop aquaculture and protect natural resources as well as motivate the public to take care of the grassland for a green environment (Central Committee of the Chinese Communist Party, 1984).
1985	The Ten Policies on Further Revitalizing the Rural Economy (关于进一步活跃农村经济的十项政策)	Regional farmer cooperatives shall engage in water conservancy and plant protection projects (State Council of PRC, 1985).
1986	Deployment on 1986 Rural Work (关于一九八六年农村工作的部署)	Within this year, competent authorities shall formulate regulations on prohibiting non-agricultural construction from occupying arable land, as well as setting up concrete measures on water and soil conservation and rural environmental protection (State Council of PRC, 1986).

Table 5 shows the relevant policy support for CEA development from 1982 to 1986. The No.1. Central Documents during this period emphasized the land problems resulting from conventional agriculture and called for a transition to a more sustainable mode that uses resources appropriately and consumes less energy so as to reduce pollution. Protecting the environment is a priority in agricultural activities. Farmers are urged to conserve water and protect plants as part of the efforts for environmental protection. The government also stipulates that the arable land area is only saved for agriculture and cannot be occupied for non-agricultural construction which releases industrial pollution.

From 1993 to 1999: During this period, relevant policies focused on the integral governance of environment in rural areas. The central government advocated and the local government experimented CEA policies. In 1993, the Ministry of Agriculture issued a report on accelerating the development of CEA, encouraging to start pilot projects from a national scale, which means that CEA implementation is officially adopted by the government (Ministry of Agriculture of PRC, 1995). In 1994, the State Council approved the report. In 1996, the development of CEA was adopted in the ninth Five Year Plan (M. Zhao, 2019).

In 2000, the State Council issued a circular on the National Ecological Environment Protection Program, which included the CEA development as an important aspect (State Council of PRC, 2001). The circular pointed out that in order to protect the ecological environment, it is imperative to construct ecological agriculture counties and strive to promote CEA on regional and provincial level where it is applicable. Safe and pollution-free agricultural products are expected from CEA (C. Gao, Li, & Qiu, 2009).

From 2000 until present: it is a period for systematic governance of environmental problems in agriculture. In 2012, the 18th National Congress of the Communist Party of China (CPC) proposed the concept of Ecological Civilisation Construction (ECC), which included CEA as one element for achieving harmony between man and nature (Hu, 2012). Since the inception of ECC, the CEA development has been continuously supported by relevant policies. Some of these policies focus on the control of use of chemical substances and the conservation of ecology. The Ministry of Finance, together with the Ministry of Agriculture, passed CEA subsidy policies since 2014. These policies are promoted to subsidise farming that uses low-toxic and low-residue pesticides and decreases the use of highly toxic pesticides (Gong & Yan, 2015). In 2015, the Ministry of Agriculture formulated requirements on the control of pesticides and chemical fertilizers, claiming that there must be no more increase in the use of pesticides and chemical fertilizers (B. Wang, 2018).

Also in 2015, the State Council issued the National Agricultural Sustainable Development Plan (2015–2030), urging to develop CEA modes that adapt to local conditions and make efficient use of resources by combining grain production with animal farming and/or aquaculture (Ministry of Agriculture of PRC, 2015). In 2016, the Ministry of Agriculture proposed that promoting CEA saves resources, conserves ecology and is environmentally friendly (Ministry of Agriculture of PRC, 2016). In 2017, the State Council issued a policy document titled Opinions on Adopting Innovative Systems and Mechanisms and Promoting Green Agricultural Development that constructs a new landscape for agricultural development that takes care of the harmony between man and nature (Luo, 2018). In the same year, the Ministry of Agriculture calls for national-level adoption of CEA that combines grain production with livestock farming. It refers to this type of CEA as circular agriculture that is efficient and environmentally-friendly (Ministry of Agriculture of PRC, 2017b). These important policies are highlighted in Table 6.

Table 6. Recent policy documents related to CEA

Year	Policy Title	Relevant Content
2015	National Agricultural Sustainable Development Plan (2015 - 2030) (全国农业可持续发展规划（2015–2030年）)	Depending on the actual problems in agricultural sustainable development, it is necessary to consider the carrying capacity of resources and the local conditions to divide different development regions. The agricultural development model shall fit the specific contexts (Ministry of Agriculture of PRC, 2015). In southern China, the focus is on decreasing the use of chemical fertilizer and improving soil fertility while developing ecological agriculture that fits the local conditions and is efficient (Ministry of Agriculture of PRC, 2015). Until 2020, the construction of ecological agriculture is expected to make substantial progress (Ministry of Agriculture of PRC, 2015).

(Cont'd.)

Table 6 cont. Recent policy documents related to CEA

Year	Policy Title	Relevant Content
2016	National Plan for Agricultural Modernization (2016-2020) (全国农业现代化规划（2016–2020年）)	Appropriately dealing with the relationship between agricultural production, increase in farmers' income, environmental protection and ecological recovery. Promoting ecological agriculture that is resource-saving, environmentally-friendly and has ecology conservation function (Ministry of Agriculture of PRC, 2016). Developing 300 demonstration counties that implement circular and ecological agriculture that combines grain production and livestock farming (Ministry of Agriculture of PRC, 2016).
2017	Opinions on Adopting Innovative Systems and Mechanisms and Promoting Green Agricultural Development (关于创新体制机制推进农业绿色发展的意见)	Ecological agriculture has profound meanings for food security, resource security and ecological security. It concerns the well-beings of the current generation and ensures sustainable development for the future generations (Ministry of Agriculture of PRC, 2017a). Under the guidance of TMT, it is imperative to implement ecological agriculture that follows the laws of nature and adopts technological innovation to save resources and improve ecological function, so as to construct a new agricultural landscape for harmony between man and nature (Ministry of Agriculture of PRC, 2017a). The basic goals of ecological agriculture are to ensure food security, green supply and increase in farmers' income (Ministry of Agriculture of PRC, 2017a).

Table 6 cont. Recent policy documents related to CEA

Year	Policy Title	Relevant Content
2017	The 2017–2020 Construction Plan for Circular Agriculture Demonstration Programs Combining Grain Production and Livestock Farming (种养结合循环农业示范工程建设规划（2017–2020年）)	The type of ecological agriculture that combines grain production and livestock farming uses animal faeces as manure for grain cultivation, and the grain cultivation provides feed for animals, which converts materials and energy between animals and crops and is therefore, a circular agriculture. Advancing such agricultural development is important for improving the efficiency of resource use and protecting the environment, as well as keeping the Two Mountains (Ministry of Agriculture of PRC, 2017b). Such ecological agriculture improves the quality of agricultural products while increasing income for farmers. When the demonstration programs are completed, it will effectively boost local employment (Ministry of Agriculture of PRC, 2017b).

Table 6 illustrates four important policies that explicitly mention the relevance of CEA to sustainable development. The National Agricultural Sustainable Development Plan points to the principle of adapting to local conditions that considers the carrying capacity of resources, which is also mentioned in the ECC (Hu, 2012). It also points out CEA construction across the country shall be divided into different development regions and there is an urgent need to decrease the use of chemical fertilizers in southern China.

In 2016, the National Plan for Agricultural Modernization mentioned the multi-functional aspects of CEA by emphasizing proper treatment of the relationship between man and nature. Stating the features of efficient production, more income for farmers, environmental protection and conservation, it proposed to develop 300 CEA demonstration counties that combine grain production and livestock farming so as to promote circular and ecological agriculture for wider implementation in the future.

In 2017, while stressing the multi-functional aspects of CEA, the concept of TMT is also used to characterize the connotation of CEA, which is to follow the laws of nature and take care of the harmony between man and nature while engaging the agricultural activities. It also mentioned the importance of technology innovation that saves resources. The Ministry of Agriculture set up three basic goals for the development of CEA, i.e., food security, green supply and more income for farmers. Also in 2017, the Ministry of Agriculture included circular agriculture as one type of CEA that combines grain production and livestock farming. In a brief description, the circular agriculture is explained and promoted as a strategy to achieve the efficient use of resources while protecting environment and increasing income for farmers, so as to contribute to keeping the Two Mountains. Since the inclusion of TMT in the 19th CPC National Congress report in 2017, the TMT and its relevance to CEA is mentioned twice in the same year from a policy level by Ministry of Agriculture.

Summing up the policy traces supporting the development of CEA, Figure 4 shows the policy focus of different time periods from 1973 to the present.

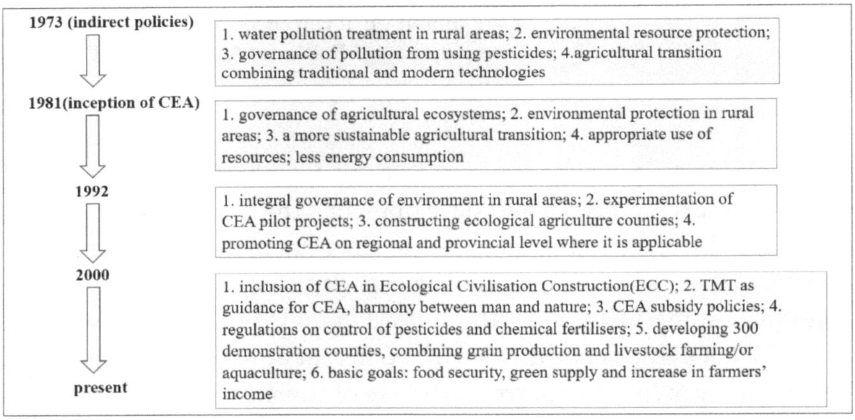

Figure 4. CEA-related policy focus from 1973 to the present

As shown from Fig. 4, the text in different colored textboxes highlights the policy focus of that specific time period. Over the past four decades, various policies paved the way for establishing and developing CEA that has been mainly implemented under a top-down approach. Recently, the goals of CEA have been set from a government level as well.

4.2 Results from the case study

The results of the case study are described in two parts—charting and interpreting the interview data, and a summary of the data interpretation.

4.2.1 Charting and interpreting the data

The 14 categories formulated after the coding process include: "advantages of CEA product," "land problem," "technologies used in CEA," "technical explanation," "farmer's perspective of CEA," "responsibility," "other farmers," "problem in CEA," "farmer's interaction with other farmers," "farmer's interaction with the interviewer," "pioneer," "farmer's agenda," "farmer's love," "changes after CEA implementation." The two separate codes that do not belong to any of these categories are "subsidy for CEA" and "exploitative production." The coded data is summarized into a total of 16 tables. Each table consists of different codes and their respective quotes. The quotes are formed using both verbatim words and summarised words, then inserted into the table with references to the page and line number. The verbatim text is underlined. Below are the framework tables (from Table 7 to Table 22).

Table 7. Codes belonging to the category "advantages of CEA product" and respective quotes from the interviewee

Code name	Quote
AA-level green food	the good thing is the test result that qualifies our rice as AA-level green food (p. 83, 139); green food certification is the result of organic certification combined with Chinese conditions. (p. 83, 142-144)
improvement of rice quality	The advantage also shows in the quality of the rice (p. 82, 115-116); the smell of authentic rice (p. 82, 118)
safe food	the rice we produced was tested 37.04 times lower than the heavy metal indicator prescribed by the green food. (p. 104, 762-763)
rice price	The wholesale price in our region is 10 RMB per 0.5kg. (p. 86, 231-232)
rice protein	Nutrients stored in solid form. The quality is guaranteed. (p. 87, 251)
economic value of crabs	high demand on the market For each mu, these crabs bring several thousand RMB to farmers. (p. 87, 262-263) (mu: Chinese unit of area, equivalent of 0.067 hectares)

As shown from Table 7, the interviewee talks about food produced from CEA that has multiple advantages. By pointing out the good quality and safety of the rice, he believes it is good to engage in such agriculture. Moreover, he mentioned the economic value brought by sale of rice and crabs, which increases the income for farmers.

Table 8. Codes belonging to the category "land problem" and respective quotes from the interviewee

Code name	Quote
soil compaction	When nitrogen goes into the soil, it mainly decomposes the SOM etc (p. 81, 61-62); SOM was largely dissolved (p. 85, 202); bad food production
increase in use of fertilizer	From 2007–2015, increase in use of fertilizer to achieve the same amount of production. (p. 80, 54-55); since 1980s, nitrogen fertilizer was applied heavily in the field. (p. 85, 199)
reason for engaging in CEA	the land has not been able to produce crops for a long time. (p. 89, 335)
land lost fertility	after use of large amount of chemical fertilizer, our land lost its fertility (pp. 80-81, 56-57); it will only answer you with the crops you have sown—it lacks nutrition. (p. 81, 68-69)
concern from the government	each Chinese President knows there is something wrong with our land and soil (p. 92, 404-406)
pollution from chemical fertilizer	soil absorbs only 13% of chemical fertilizer; 10% is taken away by water (p. 100, 661); fertilizer residue taken away by water

Table 8 illustrates a series of land problems before CEA was implemented. The interviewee talks about the land fertility loss due to heavy input of chemical fertilizer from conventional farming. The land was unable to produce crops for a long time and the government is also concerned with the land and soil, which shows a serious and continuous land problem in China. The land problem is one of the main reasons why the interviewee wants to engage in CEA. Realizing the unsustainable production by using heavy chemical fertilizers, it serves as a moment for change into a sustainable mode by converting conventional farming into CEA. Because the economic values cannot be sustained and the available resources are deteriorating, man starts to reconsider the relationship between man and nature.

Table 9. Codes belonging to the category "technologies used in CEA" and respective quotes from the interviewee

Code name	Quote
legacy from ancestors and use of biotechnology	legacy: green manure and straw-returning to the field (p. 81, 80); biotechnology: use of placeholder bacteria for plant protection
technique for cultivation	traditional cultivation method that obeys the laws of nature (p. 103, 731); use of modern technology like the plant-based material.
example of increasing microorganisms	apply human interference by adding microorganisms and fermenting it with faeces (p. 103, 750); apply it to the field with enough nutrients
the use of biotechnologies for pest control	insects eat the leaves that were sprayed with plant-based pesticide (p. 83, 128); fall dead in 10 days
rice-duck farming system	ducks reared in a free-range way (pp. 85-86, 214-215); sold at higher price than ducks from conventional farms; ducks kill insects
role of fish in CEA	fishes evacuate faeces that provide natural nutrients to the rice crop. (p. 88, 280)
rice cultivar	traditional; non-GMO; rich in nutrients; The taste is sweeter, softer and chewy. (p. 88, 298)
crab rearing	crabs feed on the freshwater snails. (p. 87, 253-254); rearing crab at right time and under right conditions, following the laws of nature (p. 87, 255)
conservation during fallow period	flip the surface soil (p. 92, 416); hairy vetch as green manure (p. 92, 418); soil protection and recovery (p. 92, 420-421); preparation for next year's plantings

Table 9 shows a series of traditional techniques and biotechnologies are used in CEA, including the use of green manure and straw-returning to the field that have been applied by ancestors. The use of biotechnologies is a feature of CEA, which helps with pest-control. When it comes to rice cultivation and crab rearing, the interviewee talks about the principle of following the laws of nature by engaging in agricultural activities at the right time and under the right conditions. Methods of land conservation are explained and reflect the care for the environment. For rice-fish-crab system, the fish faeces provide nutrients to the rice, and the crabs feed on the freshwater snails, which shows efficient use of resources within the system.

Table 10. Codes belonging to the category "technical explanation" and respective quotes from the interviewee

Code name	Quote
altitude and temperature difference	significant temperature difference results in high quality of dry matter stored in rice. (p. 86, 244)
microorgan-isms	prepare soil for microorganisms to multiply, which is the most important aspect. (p. 103, 736-737)
expert's knowledge in choosing fertilizers	recommendation from university experts; we chose five fertilizers in total and one of them is used for our trial field. (p. 90, 348-349)
long plant growth cycle	plant rice once per year (p. 88, 289); it can last up to 197 days. (p. 88, 293)
technical explanation of current CEA	insufficient carbon and other nutrients from the air; human interference by applying bio-organic fertilizer that meets the national standard NY884. (p. 84, 173)

The natural conditions of the farm are briefly explained in Table 10. The interviewee mentioned this part to clarify why the rice quality is good. The use of bio-organic fertilizer chosen by experts increases the microorganisms in the soil and provides the nutrients needed for rice growth.

Table 11. Codes belonging to the category "farmer's perspective of CEA" and respective quotes from the interviewee.

Code name	Quote
adaptation to nature	The best comes from what grows in a natural way. (p. 93, 449-450); follow the circular laws of the natural world (p. 93, 452)
first step in CEA implementation	The first step is to produce good quality rice. (p. 96, 519)
second step in CEA implementation	The second step is to improve the values the land can provide, since the land is precious and its resources are non-renewable. (p. 96, 519-521); introduce rice-fish-crab farming system. (p. 96, 523-524)
non-chemical farming	prohibiting the use of chemical fertilizer and pesticides in the crop farming system. (p. 81, 72-73)
soil has life	The land, as well as the soil, has life, although it cannot speak and you are unable to see it with naked eyes. (p. 81, 66-67)
recognition of the harmony between man and nature	The agriculture that takes care of such harmony is to be the sustainable agriculture for the future of China. (p. 85, 185-187); exploration since 2016
Taoist's view of human and nature	Man takes his law from the Earth; the Earth takes its law from Heaven; the Heaven takes its law from the Tao. The law of the Tao is its being what it is. (p. 84, 177-180); follow and adapt to laws of nature
care for the next generation	aim for development in the long run. (p. 94, 470); healthy food for next generations
contribution to Rural Revitalization	engaging in the ecological agriculture helps make farmers better off and improve ecology. (p. 96, 532)
farmer's conclusion	Good seeds, good techniques and good field are necessary. (p. 84, 157); harmony between man and nature (p. 84, 177)

Table 11 illustrates farmer's ideas about man and nature, as well as the non-chemical farming, one type of CEA. By mentioning the best comes from what grows in a natural way, he implies the principle of adaptation to nature, i.e., man adapts to nature and behaves accordingly. In his view, two steps are necessary for a successful implementation of CEA: to produce good quality rice and to increase the values the land can provide. The good quality rice is the precondition for increasing the values. By introducing the rice-fish-crab farming system, the farmer shows an example of how to increase the values. Although he did not explicitly mention what kind of values they can be, it can be deduced that economic, social, and ecological values are included because, as he put it, CEA helps farmers better off, improves ecology and produces healthy food for the next generation. The interviewee also mentioned that the land and soil have life in themselves although they cannot speak and their life is invisible, which shows the intrinsic value.

The interviewee also pointed out the need to recognize the harmony between man and nature in CEA, which is important for the future of Chinese agriculture. By quoting Taoist's view of man and nature, he mentioned the need to follow and adapt to the laws of nature. In his conclusion for CEA, three "good" are necessary, i.e., good seeds, good techniques, and good field.

Table 12. Codes belonging to the category "responsibility" and respective quotes from the interviewee

Code name	Quote
being responsible	I have to shoulder the responsibility, since I am the initiator. (p. 98, 595)
commitment to compensate loss	promise to cover the financial loss; other farmers have to consider if I am able to cover their loss as promised. (p. 99, 606-607)
cover the cost and honour the commitment	other farmers come for explanation; apply more fertilizer for crops to grow; one to two hundred thousand of RMB to cover the cost (pp. 90-91, 367-368)
harvest guarantee	guarantee that the harvest is not compromised and farmers' revenue is increased. (p. 97, 572-573)

Table 12 shows how the farmer undertakes responsibility for his promises. As the initiator of the non-chemical farming, the interviewee needed help from other farmers who were engaging in conventional farming. To persuade them to join him, he made a promise to cover the financial loss if the rice harvest is not good and does not sell a good price. When other farmers came to him and asked for explanation because the rice didn't grow with the arranged use of bio-organic fertilizers, he analysed and found the reason, i.e., the fertilizers were not enough, and covered the extra costs to apply more fertilizers. As a result, the harvest was guaranteed and farmers received more income. He honoured his commitment and won trust from other farmers.

Table 13. Codes belonging to the category "other farmers" and respective quotes from the interviewee

Code name	Quote
concerns from other farmers	the leadership of the cooperative (p. 100, 638); use of technology; good rice price
economic lure to persuade other farmers	promise of better income; I would personally cover their financial loss if the target price is not met. (p. 98, 592-593)
other farmers' feeling of CEA implementation	main reason: it brings good income (p. 99, 629); better ecosystem; farmers feel they eat healthy food. (p. 84, 161)

Table 13 explains other farmers' concerns and feelings about CEA, in the view of the interviewee. In his eyes, other farmers do not worry about how to use the technologies and are assured they will receive good income if they join the cooperative, which is the main reason why they engage in CEA. As the interviewee puts it, he will personally cover their loss if the target price is not met. Otherwise, they feel CEA improves the ecosystem and they can enjoy healthy food, which shows ecological and social benefits of CEA.

Table 14. Codes belonging to the category "problem in CEA" and respective quotes from the interviewee

Code name	Quote
loss of crabs due to distant transporta-tion	crabs purchased from another province; 45-50% of death (p. 87, 266); huge loss due to long-distance transport. (p. 87, 271)
insect hazards	Insects come back since the environment gets better. (pp. 82-83, 120-121)
insufficient fertilizer leads to problems	Recommended application of fertilizer by experts is not enough; rice did not grow. (p. 90, 362)

The current problems in CEA are illustrated in Table 14. In the rice-fish-crab farming system, a main problem is the insect hazards because the use of chemical pesticide is prohibited. As the interviewee mentioned, the biodiversity improves and insects come back when the ecosystem gets better. Another problem is the loss of crabs due to long transportation. For this farm, the crabs are purchased and transported from another province and when they reach the farm, almost half of them are dead, which leads to a question of whether the use of crabs from another place is wise and sustainable. During the first trial of applying bio-organic fertilizers to the field, the rice didn't grow because the amount of use recommended by experts was not enough.

Table 15. Codes belonging to the category "farmer's interaction with other farmers" and respective quotes from the interviewee

Code name	Quote
farmer's reflection	what can we provide to feed our next generation? Are we assured that what they will eat will be healthy? (p. 100, 635-637); good soil; good crop; good quality food; more income for farmers—believe in such ecological agriculture. (p. 103, 742)
help from experts	university experts train and motivate farmers to learn how to engage in such farming system. (p. 90, 338-339)
no voluntary participation	No one volunteered to participate. (p. 98, 585)
sharing dividends	distribution of dividends based on how much rice one contributes to the cooperative. (p. 99, 619-620)
trust of other farmers	As a native, I am trustworthy. (p. 89, 333-334); I won their trust because of my interpersonal relationships and my moral qualities as a person. (p. 98, 600-602)
happiness of harvest	we are happy to see the crops are ripe and without any diseases. It's a happy feeling. (p. 89, 323-324); increase in food production and income.

Table 15 shows the activities that involve all farmers in CEA. The interviewee had to first convince other farmers to join him for this type of farming, and he did it because he keeps good relationships with other farmers and he is a native whose ancestors also lived there. As the leader of the cooperative, and together with university experts, the interviewee holds regular training sessions with other farmers to teach them how to use technologies. The interviewee asked two questions in front of other farmers: what they can provide for next generations, and if they are assured the food will be healthy. In the end, the interviewee stressed that good quality food, good soil, and good income for farmers can make them believe in such ecological agriculture. This is also partly proved by the non-voluntary participation from other farmers at the beginning of the CEA. Other farmers are assured they can get good income because they share the dividends based on how much rice they contribute to the cooperative; therefore, it is also a fair distribution of income. When it comes to harvest, every farmer is happy because they know it will guarantee good food production and income.

Table 16. Codes belonging to the category "farmer's interaction with the interviewer" and respective quotes from the interviewee

Code name	Quote
farmer's invitation	you are welcome to come to our field and see it yourself. (p. 85, 212)
farmer's honesty	if it is not because you are also from Xichang, I will not tell you the secret at all. (p. 82, 109-110); secret of increase in production.
farmer's generosity	The day when your dad visited me, I gave him a few packs of rice as gift. (p. 86, 236-237)
farmer's gratitude	I thank you for offering this opportunity to let me introduce this type of CEA. (p. 86, 240-241)
farmer sharing a lesson	Only with practice can the research draw a conclusion. (p. 90, 353-354); research is critical. Avoid entering a wrong direction (p. 90, 355)
distracted farmer	Continue talking, I hear you. (p. 91, 387) (At this moment he did not show up in the screen as he was looking for something.)
reluctance to recommend other farmers for interview	Well… (p. 104, 774); (pondering for a few seconds) I am confident that I answer better than them. (p. 104, 776-777); we are all very busy with production. (p. 104, 781-782)

Table 16 shows the conversations that build up the connection between the interviewee and the interviewer. As mentioned in the methodology section, the interviewee was identified during an actual farm visit by two local contacts of the interviewee; therefore, the farmer already had an initial and indirect connection with the interviewer. During the interview, the farmer was willing to share his knowledge. He shared a lesson from the insufficient application of fertilizer recommended by the experts and concluded that only with practice can the research draw a conclusion. He was honest about sharing the secret of the increase in food production because the interviewer came from the same city where he is and he was generous to gift the visitors a few packs of rice. Moreover, he was grateful for participating this interview and invited the interviewer to visit his farm. The farmer was distracted as he did not appear in the screen for a few seconds but he was still paying attention to the interviewer by asking him to continue. In the end of the interview, the farmer showed reluctance as the interviewer asked him to recommended other farmers for interview. However, since the interview took place when it was close to the harvest season, it was quite difficult for the farmer to consent to this request immediately.

Table 17. Codes belonging to the category "pioneer" and respective quotes from the interviewee

Code name	Quote
being a pioneer	there must be someone who starts doing this, and I would like to start this. (p. 94, 479-480)
courage to try something new	dare to give it a try; it has not been always easy and straightforward. (pp. 95-96, 516-517)
determination to implement CEA	after a few years of such farming practice, we came to the conclusion that the future of Chinese agriculture should promote it, regardless of support from the government. (p. 83, 147-149); initiator of non-chemical farming.
farmer's gratefulness	now there are a large group of people supporting me, and I am sincerely grateful to them. (p. 99, 621-623)
experienced farmer	30 years of farming experience. (p. 80, 38-39)

Table 17 shows the qualities of the interviewee who becomes the first one to implement non-chemical farming. As an experienced farmer, he dares to try this type of CEA partly because he wants to change the situation of the land impacted by conventional farming and he is willing to be the proactive one. He also demonstrated this courage continuously when there were problems during the process, as he put it, "the road is not always easy and straightforward." When his efforts paid off, he is determined to implement CEA no matter whether this type of farming receives actual support from the government or not. As a pioneer, he is now backed by many people to whom he is grateful.

Table 18. Codes belonging to the category "farmer's agenda" and respective quotes from the interviewee

Code name	Quote
calling for sustainable change in agriculture	the future of agriculture in China cannot continue with the reliance on heavy chemical fertilizers. (p. 82, 103-104); otherwise we all become sick man of Asia. (p. 82, 105)
healthy food for future generation and economy	It is important for me to protect the land that produces healthy food for our future generations. (p. 82, 97-98); contribute to economy
merits of CEA	This farming is a project that not only benefits current people's livelihood but also takes care of the future generations. (p. 100, 642-644)
transition of CEA	transitioning towards a model that aligns with governmental policies and adapts to the nature. (p. 102, 718-719)
rice pilot project	rice agriculture pilot project with the government (p. 85, 191-192); healthy food for future generations.

Table 18 exhibits the farmer's plan, hopes, and the appeal of CEA. While he calls for sustainable change in agriculture, he mentioned the need to stop the use of heavy fertilizers—otherwise the food will not be safe and of good quality, which leads to a similar situation of sick man of Asia. He believes it is important to produce healthy food for next generations which also contributes to the economy. He stressed the point that CEA benefits both the current and the future generations as he reports to government officials from the State Council. As for his plan, he wishes to continue implementing CEA and transition towards a model whose development responds to policies and is shaped together with the government.

Table 19. Codes belonging to the category "farmer's love" and respective quotes from the interviewee

Code name	Quote
dedication to agriculture	no difference between work and leisure time. (p. 102, 703-704); When I think about the plants in midnight, I go there. (p. 102, 704-705); summarize learning and share with farmers; for me there is no such thing as work. I am always there, day and night... (p. 102, 710-711)
love for the land	one has to feel attached to the land, love the land, and proactively learn from experience. (p. 94, 463-465); I can simply say that I love agriculture and I love the land. (p. 95, 504-505); I just felt the need to plant the rice in a good way. (p. 95, 510); I am in the field almost at any time because I love agriculture and I feel like going there. (p. 102, 701-703)
meaning and value	For me the main thing is about the meaning and value as an individual. (p. 94, 476-477); We come to the world with nothing on our hands, but we want to contribute something meaningful to the society before we pass away. (p. 94, 477-479)

Table 19 shows the farmer's love for agriculture. Specifically, the interviewee mentioned that there is no difference between work and leisure because he is always there taking care of the farm. In order to learn how to grow the rice well, he goes to the farm at midnight and observe the growth so that he can share it with other farmers. He feels it is his obligation to protect the land and devote himself to it. This sentiment is reinforced multiple times by showing his love for agriculture. When it comes to why he devotes himself to agriculture, he expressed that it is about how a person perceives the meaning and value as an individual. As a farmer, he feels he needs to contribute something meaningful by engaging in CEA before he passes away.

Table 20. Codes belonging to the category "changes after CEA implementation" and respective quotes from the interviewee

Code name	Quote
recovery of vitality and life	fertilised soil; deeper arable layer; As the plants grow up, they appear strong and not sick. (p. 82, 90)
recovery of ecosystem as incentive	animals and various fishes come back, so it is a good sign. (p. 82, 93); I believe in this type of farming. (p. 82, 93-94); communicate with fellows in conferences
increase in production	compared to conventional farming, our farming system is able to increase 100kg per mu. (p. 82, 113-114); (mu: Chinese unit of area, equivalent of 0.067 hectares)

As shown from Table 20, positive changes took place after the CEA implementation. The soil is fertilised, no longer suffering from soil compaction, and plants grow stronger, which leads to increased rice production and contributes to food security. The improved environmental conditions also bring back living animals. With the positive changes, the interviewer communicates it with his fellows in agricultural conferences. These positive changes reflect economic, social and ecological benefits of CEA.

Table 21. The code "exploitative production" and respective quotes from the interviewee

Code name	Quote
exploitative production	You are talking about the exploitative production, which in my opinion is wrong and will not work in the long run. (p. 93, 445-446)

Table 21 shows the interviewee's comment on interviewer's explanation of TMT. The interviewer first introduced the concept of TMT and exemplifies it with non-chemical farming that prohibits the use of chemical fertilizers and pesticides. As a response, the interviewee shortly mentioned it is wrong to go for exploitative production that increasingly uses chemical fertilizer to produce more food at the cost of the environment and the soil because it is not sustainable and won't work in the long run.

Table 22. The code "subsidy for CEA" and respective quotes from the interviewee

Code name	Quote
subsidy for CEA	Significant incentives from policies and projects; Specifically, for use of bio-organic fertilizer, each mu is subsidized by dozens of RMB. (p. 91, 393-394); The bio-organic fertilizer we are using for the field costs at least 600RMB per pack, and we get more than 100 RMB for the subsidy. (p. 92, 396-398); It encourages you to use such fertilizers. (p. 92, 403-404); I knew the subsidy after I started this farming system. (p. 96, 539); I have already done it for two years before there was such subsidy. (p. 97, 552-553); This subsidy comes from a national-level project that indicated the need to decrease the use of chemical pesticides and fertilizer, and to increase the use of bio-organic fertilizer. (p. 97, 560-562) (mu: Chinese unit of area, equivalent of 0.067 hectares)

Table 22 illustrates the type of subsidy that the farm receives for implementing CEA. It is a national subsidy that encourages use of bio-organic fertilizer. According to the interviewee, for each pack of the bio-organic fertilizer, the government will subsidize over 100 RMB as an incentive for use. The interviewee came to know such subsidy two years after he implemented non-chemical farming.

4.2.2 Summary of the data interpretation

In relation to the research questions, a summary is made in the following:

As expressed in CEA's mainstream definition and relevant CEA policies, the economic, social and ecological benefits are reflected multiple times during the interview. Specifically, extra income from rearing crabs, safe and healthy food (see Table 7); helping farmers be better off, improving ecology and producing healthy food for next generations (see Table 11&13); guarantee of good quality food, good soil and good income leading to belief in CEA (see Table 15); CEA's contribution to food security as a social benefit(see Table 20). These benefits are all derived from nature's benefits and therefore belong to instrumental value.

As revealed in CEA-related policies, characteristics of CEA are identified under the categories of "technologies used in CEA" and "technical explanation." They are: combination of traditional techniques and modern technologies (specifically, biotechnologies), conservation of land, environmental protection, adaption to local conditions, and efficient use of resources through combing aquaculture with grain production (see Tables 9 & 10).

The intrinsic value is mentioned one time during the interview when the interviewee acknowledged that "soil has its own life" (see Table 11).

Another finding is the presence of relational value. As shown in the interview, harmony between man and nature is pursued in CEA, i.e., finding a balance between agricultural production and land conservation/environmental protection (see Table 11); also during the harvest season, relational value is reflected from farmers feeling happy viewing the healthy and ripe crops and that their hard works are paid off (see Table 15) ; the farmer's love for engaging in agriculture with continuous dedication and his need to contribute something meaningful from agriculture to the society is a realisation of fulfilled life, therefore, it belongs to relational value (see Table 19).

Held value is also identified as an influencing factor in CEA implementation; the interview found that responsibility (to honor promises), courage (as a pioneer), determination (to continue CEA), confidence (as an experienced farmer) and obligation (to protect land and ensure healthy food for future generations) all played a role (see Tables 12, 17 & 18) because the interviewee demonstrated such qualities necessary for success of CEA implementation. Government played a role as well, i.e., formulating policies and providing financial support for purchasing bio-organic fertilizers (see Table 22). Other than that, social justice is achieved in this case study as farmers receive a fair share of the dividends based on how much they contribute to the cooperative (see Table 15).

5. Discussion

5.1 Discussion of results in relation to research questions

The literature review and the case study both address the research questions in varying degrees. Particularly, the literature review answers mainly the first two questions and the case study answers mainly the third one. Given that the second and third questions are interrelated, for the structure of the discussion part, they will be discussed together.

First question: "What is the concept of ecological agriculture as expressed in current policies in China?"

The literature review on how the concept of CEA is formulated and developed shows that the CEA is taking shape both from a conceptual and policy level over time. From its inception in 1981 to recent years, the content of CEA was continuously enriched from producing food while obeying the laws of nature, to including efficient use of resources with the help of technologies and increasing income for farmers. Many scholars contributed to the development of the concept but the theoretical debate remains on what exactly is CEA. Some scholars believe that since the definition of CEA is too broad, and a lot of agricultural practices are related to it, they all can be consider ecological agriculture (Qi et al., 2021; Tan, 2019). Others argue that CEA is a social movement that can be realized at grassroots level and that developing CEA serves as a chance to motivate social groups to support sustainable agriculture (Shi, 2003). However, examples of such social movement were not provided and the potential of such bottom-up approaches are not yet tested to see how government recognises it. Contradictory to this is the framing of CEA as a top-down approach where policy makers set up CEA demonstration counties on a small scale and encourage wider application when it is proved successful (Ministry of Agriculture of PRC, 1995, 2016, 2017b).

Moreover, some scholars addressed CEA in their research as farming without chemical materials (Shao et al., 2019), which corresponds to the non-chemical farming in the present case study. The development of CEA theory inspired different modes in practice. Scholars summarized four common rules for a successful CEA implementation, which are condensed in four adjectives, being circular, green , low-carbon and efficient (Y. Zhang et al., 2015). These four rules are also embodied from the case study where the characteristics of CEA are identified through the farmer's initiation of non-chemical farming and the combination of aquaculture with grain production (see Tables 9 & 10).

As CEA develops, the relationship between man and nature is addressed in policies. On a national strategic level, CEA was promoted as one of the sustainable approaches to achieving the harmony between man and nature since 2012 (Hu, 2012). The relevance of TMT in CEA development is also briefly mentioned in government agendas (Ministry of Agriculture of PRC, 2017b, 2017a).

The broad conceptual debate on CEA also leads to another question, i.e., how is CEA differentiated from other types of sustainable agriculture. This is also revealed in policy documents that address CEA with various features through the sustainability transition. As a model of agricultural development, CEA is evolving gradually and its relevant policies show a trend towards dealing with different stages of issues related with rural area, farmers and agriculture (M. Zhao, 2019). From the compilation of policies (see Tables 5 & 6), the focus of policies changed from protecting environment and governance of ecological problems in rural areas, to systematic governance of environmental problems in agriculture. In this process, the different features of CEA mentioned in the policies included: adapting to local conditions, circular, efficient, resource-saving, and environmentally-friendly.

To sum up, the expression of CEA in policies has evloved over time. It is expressed mostly in changing characteristics that reflect the pressing issues in agriculture over a certain period of time and is still evolving. Since the inception of CEA, relevant policies emphasize the combination of tradional and modern technologies to save resources and achieve production increase and environmental protection (W. Li et al., 2010). Pilot projects started in selected places across the country (Ministry of Agriculture of PRC, 1995). Then, wider implementation of CEA on regional and provincial level is encouraged (State Council of PRC, 2001).

Since 2010, more concrete CEA policies stipulated standards on control of pesticides and chemical fertilizers (B. Wang, 2018), while urging to develop efficient CEA modes that combine grain production with animal farming and/or aquaculture (Ministry of Agriculture of PRC, 2015, 2017a); Later on, ECC included CEA as one element towards achieving harmony between man and nature (Hu, 2012), following this trend, a further requirement of CEA considers the carrying capacity of resources and the local conditions. Specifically, it states to decrease the use of chemical fertilizer and improve soil fertility (Ministry of Agriculture of PRC, 2015). The meaning of CEA is also refined during this period, which includes ensuring food security, resource security, ecological security, and increase in farmer's income, although these aspects are mentioned only briefly (Ministry of Agriculture of PRC, 2017a).

Over the last few years, policies start to mention TMT as the guiding framework for CEA. "Under the guidance of TMT, it is imperative to implement ecological agriculture that follows the laws of nature and adopts technological innovation to save resources and improve ecological function, so as to construct a new agricultural landscape for harmony between man and nature" (Ministry of Agriculture of PRC, 2017a). Thus, advancing CEA implementation is a necessary step towards efficient resource use and environmental protection, i.e., keeping both the Two Mountains (Ministry of Agriculture of PRC, 2017b).

Second question: "What are the values in CEA?" Third question: "How do values affect the ecological agriculture implementation in China?"

The ethical thinking in CEA is inspired by traditional agriculture that applies the principles of time suitability, land suitability, and carrying capacity of land. The time suitability focuses on engaging in agricultural activities according to the right time and following the laws of nature. The land suitability corresponds to the idea of adapting to local conditions which is specified in CEA since its inception. The carrying capacity of land focuses on conservation of land and is mentioned in CEA-related concepts (ECC for example) explicitly since 2012 (Gu et al., 2013). The ethical thinking is also explicitly found in the case study as the interviewee explained the technologies applied in CEA: "rearing crabs at the right time and conditions, following the laws of nature," and the use of green manure for conservation of land during the fallow period (see Table 9); by quoting Taoists' view on nature "the law of the Tao is its being what it is," he mentioned the importance of following the laws of nature, i.e., do appropriate things at the right time and under the right conditions, which connects to the ethical thinking of time suitability and land suitability (see Table11). The ethical thinking and farmer's reflection points to adaptation to nature, which is a step towards harmony between man and nature.

The ethical thinking inherited from traditional to ecological agriculture also reflects how man perceives himself in relation to nature. The development of CEA sees a change from man dominating nature to being a steward of nature. By pointing out the benefits that nature can offer and the fact that nature is a living system that has its own life, CEA recognizes both the instrumental and intrinsic values of agriculture (Ye & Luo, 1987). As shown from the concept as well as from CEA-related policies, economic, social, and ecological benefits are pursued, which directs to the instrumental value. This is also proved from the case study where the interviewee mentioned multiple times how CEA increased income for farmers, recovered the ecosystem and contributed to food security, including: the good quality of rice and the economic value from selling rice and crabs that increase income for farmers (see Table 7), improved biodiversity, stronger growth of plants, fertilised soil as well as increased production (see Table 20). The intrinsic value is significantly less addressed both in policies and in CEA implementation, as the interviewee only mentioned it one time with the quote "The land, as well as the soil, has life, although it cannot speak and you are unable to see it with naked eyes" (see Table 11). The rare recognition of intrinsic values in polices might be partly attributed to the fact that anthropocentric values dominate the current political discourse and the awareness on environment protection is out of self-interest, i.e., the ecological benefits for human, rather than reflecting on ecological justice.

One important finding, though, is the relational value in CEA implementation which is not identified from the literature review. In the case study, the interviewee explicitly expressed his love for agriculture multiple times. As an experienced farmer, he continues to actively learn and share knowledge with other farmers because he feels it is his obligation: "I just felt the need to plant the rice in a good way" (see Table 19). His devotion to agriculture comes from how he perceives meaning and value as an individual. He feels the need to contribute meaningful things to the society (see Table 19). This type of devotion and feeling satisfied with what he is doing with life is directly related to good life denoted by relational value.

The case study also found that economic value played a significant role during the initial stage of CEA implementation. According to the interviewee, as the initiator of non-chemical farming, he needed manpower to engage in this new type of CEA, and to achieve this, he used economic lure to convince other farmers by promising them good harvest and better income (see Tables 12 & 13). As the interviewee mentioned the high market demand for crabs, the introduction of carb in the rice-fish farming system might be highly money-driven (see Table 7). It is until later when other farmers enjoyed healthy food and better environment conditions that they start to pay attention to social and ecological benefits (see Table 13). This is also supported by relevant literature that environmentally-friendly activities will be adopted only when they make economic sense to farmers (Shi, 2003).

Other than that, a set of personal qualities including courage, determination, confidence, responsibility, and obligation played a central role in CEA implementation. These qualities belong to held values. Being a pioneer of non-chemical farming, the interviewee demonstrated his confidence, courage and determination to implement CEA (see Table 17) and responsibility in compensating the loss as promised for his colleagues (see Table 12). His personal motivation is also revealed in the call for sustainable change in agriculture. In his view, the food produced from conventional agriculture cannot continue, otherwise the people will become sick man of Asia (see Table 18). This view is reinforced by his obligation to protect the land and provide healthy food for future generations.

Another important point is the role of policies in CEA implementation, which is evidenced both from the literature review and the case study. During the interview, the conversation on subsidy for CEA indicates how important the government perceives CEA. With the national policy that provides certain subsidies for purchasing bio-organic fertilizers, it can be seen that the government is motivating the implementation of CEA that decreases the use of chemicals and increases the use of bio-organic materials (see Table 22). This is supported by literature review on policies that stipulate the control of pesticides and chemical fertilizers as well as improvement of soil fertility (Ministry of Agriculture of PRC, 2015; B. Wang, 2018). Interestingly, when envisioning the plan for CEA, the interviewee pointed out a transition toward a model that not only adapts to nature, but also aligns with governmental policy (see Table 18), which implies the interference of government in the CEA development. However, he also expressed in the interview that he would implement CEA even if there is no support from the government, which shows his personal belief in CEA plays a stronger role.

A separate section is created here for reflection on two points that do not directly relate to the research questions but are important for further study, in order to better understand CEA modes and the poorly addressed connection between CEA, TMT, and the relevant value dimension. The first point is regarding the sustainability of CEA. The type of CEA as implemented in the case study does not come without a problem. Except for insect hazards which are inevitable, the loss of crabs due to long distance transportation raises the question of whether this is a sustainable way to introduce crabs from other regions instead of using local ones (see Table 14).

The second point is on TMT in relation to CEA. Since the concept of TMT was brought up by the central government, the connotation of CEA has been refined. TMT aims to achieve harmony between man and nature through sustainable development, to which CEA contributes. Since TMT recognizes that man and nature have their own values and that man relies on nature for survival, man shall adapt to nature as he gains benefits from it. TMT suggests that man and nature are both the subject, that man and nature have their own values, and that not only man relies on nature, but every living-being relies on nature. Environment protection and economic growth are not absolutely opposed to each other, but the key lies in how man reconciles the relationship. Clear waters and green mountains are not only natural resources, but also social and economic fortunes. The economic, social and ecological benefits provided by nature improve people's livelihood and well-being. As the living standards improve, people show increasing care about environmental quality. Blue sky, fresh air, and clean waters ... living in such environment is desirable and improves the quality of life. Good environment conditions are required for the good quality of life, which cannot be replaced by economic values. There is no happiness when one has the economic benefit but lives in an unsustainable environment where air pollution exists and drinking water is contaminated. In this sense, relational value is recognised, though it has to be interpreted and is context-dependent.

5.2 Reflection on the methods

For the literature review, the inferences are drawn based on English and Chinese documents mainly selected from two databases, the Scopus and CNKI. As mentioned in Table 1, the combinations of keywords are used for searching the journal papers, some of which are inaccessible. Therefore, it might influence the result of the literature review. Another limitation might be the quality of the documents. Given the poor scholarship on this topic, various types of documents were selected, including papers that are not peer-reviewed, however, the inclusion of such papers is necessary because they do contribute to an academic debate on the role of values in sustainability transition. For a further and comprehensive look into how the concept of CEA is formulated, it makes sense to conduct a scoping or systematic literature review.

For the case study, the transcription, translation and coding processes are done by author himself because of the limited time frame and resources. However, this might be adding more subjectivity to the analysis, thus influencing the data interpretation to certain extent. In order to take a more objective approach, future research shall consider obtaining information from peer reviewers for triangulation.

As the interviewer, the author did not follow up when the interviewee failed to answer the question on if his farm was monoculture before CEA implementation (see Question 3 in Table 4). He might not get the interviewer's question and thought of something else therefore kept explaining the current farming system. However, the author chose to let him continue without reminding him or coming back to the question at a later stage. When probing information on TMT and its relevance to CEA, the interviewee only made a short comment and the interviewer did not catch the chance to dig further into the topic by asking him directly how he thought about it in detail (see Table 21). The reason for the short comment might be that the interviewer made several long sentences in a row and used some technical terms such as "sustainable transition," which is quite challenging for the farmer to comprehend. Another reason might be that the interviewer did not continue going back to the topic of TMT and trying to explain it in simpler words. In future research, special attention shall be paid to properly asking questions in a way that can elicit meaningful answers.

During the interview, the interviewee was quite cooperative and responded positively almost to all questions, except when he was asked to recommend a few other farmers for interview (see Table 16). This hindered the attempt to get more information from different farmers and, due to this reason, the interviewer was unable to look for evidence on the credibility of the interviewee's comments on how other farmers perceive CEA and how they actually joined the cooperative. Moreover, seven interruptions occurred in the interview and they all came from the interviewee's end, one of which ended the interviewer's thought on probing more information about the harvest, thus influenced the quality of the data. More information is available in Table 23 in the appendix.

A separate section is made here to describe the interviewer's attempt to look for policy makers. Originally, the author planned to also interview a couple of policy makers for their insights on CEA in relation to TMT. Since the TMT was proposed and implemented from a policy level, it makes sense to see how local policy makers view the role of TMT in CEA implementation. Therefore, the author resorted to the official website of Liangshan Yi Autonomous Prefecture Bureau of Agriculture and Rural Affairs (http://nyncj.lsz.gov.cn/) and contacted local media for support. To do this, the author tried to search for contacts from the government website and e-mailed them. However, as of the completion of this thesis, no response was received. The author also tried to ask the interviewee for recommendation of policy makers as he frequently attends agricultural conference in the region. The good news is that he indeed found some contacts and even introduced the theme of the topic to them; however, according to the interviewee's information, they all rejected participating in an interview because they were cautious and sensitive about attending interviews with students from foreign universities. The interviewee also implied that they would take the interview if the author was doing research for a Chinese university. Nevertheless, the author didn't have the chance to contact these policy makers directly and therefore cannot make judgements on that. Still, for future research, it is worthwhile to directly contact the officials and even visit them for clarifying the nature of the research and build up a certain degree of trust, which was not possible for the present research due to the long distance, limited timeframe, and COVID pandemic regulations.

Although only one interview was conducted for the case study, which cannot be representative, it is rich in useful and interesting information that contributes to the debate on how values play a role in the CEA implementation and therefore, contributes to research on the role of values in sustainability transition. Particularly, the presence of relational value in CEA implementation is an interesting finding that has not yet been addressed in Chinese literature.

6. Conclusion (with outlook)

Values are important instruments to understand ethical foundations of actions and behaviours. They provide unique look into complex issues such as sustainability transition. Given that values are constructed in various ways and across different societies and cultures, it is essential to classify and differentiate different value systems in order to provide a common ground for discussion. Such debate is not only useful for scholars to contribute to theoretical development of certain concepts, but also meaningful for policy makers by guiding directions in decision-making processes. The construction of CEA and its implementation is an example for exploring how values play a role in the sustainable development. The identification of instrumental and intrinsic values in literature review and that of relational and held values in case study show that values can be interrelated in specific contexts and that they influence conducts and actions in varying degrees. Some values might arise earlier and play a pre-dominant role (such is the case for economic values in CEA implementation), other values appear later and become intertwined with each other (such as held values and relational values). Nevertheless, the potential of CEA in its value dimension is not fully explored. For example, many scholars mentioned the multi-functional aspects of CEA, but they rarely critically addressed certain values, such as cultural and spiritual ones (Zhu, 2009). Therefore, future research is recommended to take this as an opportunity to enrich the various aspects of CEA values, with which a better understanding of man-nature relationship can be realized.

After nearly 40 years of development, the concept of CEA has been enriched both in content and in meaning. The content of CEA includes the combination of traditional agricultural wisdom with modern innovative technologies to achieve economic, social and ecological benefits while maintaining the carrying capacity of resources and contributing to increase in farmers' income. The meanings of CEA are shaped both by traditional wisdom and political rhetoric that deals with the relationship between man and nature. Specifically, the view of Sancai, the principles of land suitability, time suitability, the carrying capacity of land, as well as the TMT contribute to respecting and adapting to the laws of nature, so that man and nature stay in dynamic harmony.

Although theories can inform policy-making, it can also create obstacles when it is unclear. This is true of CEA. Given that the debate on CEA from both conceptual and policy levels is increasing, so far there is still no basic law of CEA that is made from the national level, which creates a unique challenge to construct a definitive understanding of CEA and relevant policy-making. For the same reason, there is a lack of practical measures to address the obstacles of CEA development because the regulations applied so far are fragmented, resulting in gaps between policy rhetoric and practical implementation (Shi & Gill, 2005). Therefore, concrete steps towards transition into CEA need to be systematically operationalised (Luo, 2015). Particularly, the top-down approaches where the state and market dominate the CEA development can be complemented by bottom-up initiatives from community level (Shi, 2003). The knowledge from different perspectives, especially from farmers as producers and citizens as consumers, can be of great help in this regard.

References

Baxter, B. (2004). *A Theory of Ecological Justice* (1st ed.). London: Routledge.

Bieling, C., Eser, U., & Plieninger, T. (2020). Towards a better understanding of values in sustainability transformations: ethical perspectives on landscape stewardship. *Ecosystems and People, 16*(1), 188–196. https://doi.org/10.1080/26395916.2020.1786165

Central Committee of the Chinese Communist Party. (1983). 当前农村经济政策的若干问题(The Several Issues Concerning Current Rural Economic Policies). Retrieved November 12, 2021, from http://www.reformdata.org/1983/0102/7467.shtml

Central Committee of the Chinese Communist Party. (1984). 关于一九八四年农村工作的通知(the Notice on 1984 Rural Work). Retrieved November 12, 2021, from http://www.reformdata.org/1984/0101/1836.shtml

Chan, K. M., Gould, R. K., & Pascual, U. (2018). Editorial overview: Relational values: what are they, and what's the fuss about? *Current Opinion in Environmental Sustainability, 35*, A1–A7. https://doi.org/10.1016/j.cosust.2018.11.003

Cheng, X., Han, C., & Donald C., T. (1992). Sustainable Agricultural Development in China. *World Development, 20*(8), 1127–1144. https://doi.org/10.1016/0305-750X(92)90005-G

ChinaDaily. (2017). 党的十九大报告双语全文(Bilingual Report of 19th National Congress of CPC). Retrieved June 12, 2021, from http://www.chinadaily.com.cn/interface/flipboard/1142846/2017-11-06/cd_34188086.html

Corbin, J. M., & Strauss, A. (1990). Grounded theory research: Procedures, canons, and evaluative criteria. *Qualitative Sociology, 13*(1), 3–21. https://doi.org/10.1007/BF00988593

Díaz, S., Demissew, S., Carabias, J., Joly, C., Lonsdale, M., Ash, N., … Zlatanova, D. (2015). The IPBES conceptual framework - connecting nature and people. *Current Opinion in Environmental Sustainability, 14*, 1–16. https://doi.org/10.1016/j.cosust.2014.11.002

Dong, S., Ren, J., Fang, X., Yang, M., Zhang, J., & Qi, B. (2018). 种植业的农业伦理学之度(Degree of Ethics in Crop Plantation of Agricultural Sector). *Pratacultural Science, 35*(10), 2299–2305.

Gao, C., Li, B., & Qiu, J. (2009). 中国生态农业组织管理研究(Research on Organizational Management of Ecological Agriculture in China). *Chinese Agricultural Science Bulletin, 25*(24), 392–397.

Gao, S. (2019). *"绿水青山就是金山银山"理念研究(Research on the concept that lucid waters and lush mountains are invaluable).* Party School of Heilongjiang Provincial Committee of C.P.C.

Ge, J., & Yuan, M. (2009). 道教与中国传统农业关系略考(On the relationship between agriculture and daoism). *Journal of Fujian Normal University (Philo Sophy and Social Sciences Edition), 156*(3), 137–142.

Gong, Q., & Yan, G. (2015). 中国生态农业发展的进展、问题与展望(The Progress, Problem and Prospect of Chinese Ecological Agriculture). *Modern Economic Research,* (9), 63–67. https://doi.org/10.13891/j.cnki.mer.2015.09.013

Gu, S., Hu, Y., & Zhou, H. (2013). 生态文明建设的科学内涵与基本路径(Ecological civilization construction: scientific connotation and basic paths). *Resources Science, 35*(1), 2–13.

Guo, Z. (2015). 中国生态农业的产业发展(The industrial development of Chinese ecological agriculture). *Theoritical Horizon,* (10), 29–33.

Hao, F., & Zhang, F. (2014). 试论古代"三才"理论生态观与现代生态农业发展(The Ecological Concept of Universe-Earth-Human and Modern Eco-agriculture Development). *Journal of Anhui Agricultural Science, 42*(14), 4556–4557.

Hu, J. (2012). *胡锦涛在中国共产党第十八次全国代表大会上的报告(President HU Jintao's report at the 18th National Congress of the Communist Party of China).*

Huang, M. (2016). 中华人民共和国政区大典·四川省卷(Compendium of Administrative Regions of the People's Republic of China: Sichuan Province). In *China Society Press.*

Keeble, B. R. (1988). The Brundtland Report: "Our Common Future."

Medicine and War, 4(1), 17–25. https://doi.org/10.1080/07488008808408783

Kiley-Worthington, M. (1981). Ecological agriculture. what it is and how it works. *Agriculture and Environment, 6*, 349–381.

Lai, Y., & Wang, J. (2019). 生态农业伦理的价值与事实之合(The Combination of the Value and Fact of Ecological Agriculture Ethics). *Journal of Central South University of Forestry & Technology (Social Sciences), 13*(6), 15–20. https://doi.org/10.14067/j.cnki.1673-9272.2019.06.003

Li, G. (2009). Thought and Practice of Sustainable Development in Chinese Traditional Agriculture. *China Agricultural Economic Review, 1*(1), 97–109. https://doi.org/10.1108/17561370910915393

Li, W., Liu, M., & Min, Q. (2010). 中国生态农业的发展与展望(Progress and perspectives of China's ecological agriculture). *Resources Science, 32*(6), 1015–1021.

Lin, W., & Chen, T. (2019). 中国农业的生态化转型与发展生态农业新视野(Transition of Agricultural Systems to Ecologicalizaton and New Vision of Modern Eco-agriculture Development in China). *Chinese Journal of Eco-Agriculture, 27*(2), 169–176. https://doi.org/10.13930/j.cnki.cjea.181056

Liu, P., Huang, Y., Zhao, Y., Qiu, X., Wang, Y., Luo, X., & Weng, B. (2017). 高效生态农业转型升级的战略思考与技术对策研究(Strategy Consideration and Technical Countermeasures Research for the Transformation of High Efficient Ecological Agriculture). *Ecological Economy, 33*(8), 105–111.

Lu, J., & Li, X. (2006). Review of rice-fish-farming systems in China – One of the Globally Important Ingenious Agricultural Heritage Systems (GIAHS). *Aquaculture, 260*(1–4), 106–113. https://doi.org/10.1016/j.aquaculture.2006.05.059

Luo, S. (2015). 构建我国农业生态转型的政策法规体系(constructing policy and regulation system for ecological transition of Chinese agriculture). *Acta Ecologica Sinica, 35*(6), 2020–2027. Retrieved from doi: 10.5846 / stxb201412282591

Luo, S. (2018). 中国生态农业制度的构建(Setting up Policy System for Eco-agriculture in China). *Chinese Journal of Eco-Agriculture, 26*(5), 759–770. https://doi.org/10.13930/j.cnki.cjea.180210

Ma, S., & Li, S. (1987). *中国的农业生态工程(Agroecological Engineering in China)* (1st ed.). Beijing: Science Press.

Markard, J., Raven, R., & Truffer, B. (2012). Sustainability transitions: An emerging field of research and its prospects. *Research Policy, 41*(6), 955–967. https://doi.org/10.1016/j.respol.2012.02.013

Ministry of Agriculture of PRC. (1995). 关于加快发展生态农业的报告(Report on Accelerating the Development of Eco-agriculture). *Eco-Agriculture Research, 3*(1), 1–4.

Ministry of Agriculture of PRC. (2015). 全国农业可持续发展规划(2015–2030年)(National Agricultural Sustainable Development Plan (2015–2030)). *Gazette of the Ministry of Agriculture of the People's Republic of China, 141*(6), 4–16.

Ministry of Agriculture of PRC. (2016). 全国农业现代化规划(2016～2020年)(National Plan for Agricultural Modernization (2016-2020). *Gazette of the Ministry of Agriculture of the People's Republic of China, 158*(11), 4–19.

Ministry of Agriculture of PRC. (2017a). 关于创新体制机制推进农业绿色发展的意见(Opinions on Adopting Innovative Systems and Mechanisms and Promoting Green Agricultural Development). *Gazette of the Ministry of Agriculture of the People's Republic of China, 169*(10), 4–9.

Ministry of Agriculture of PRC. (2017b). 种养结合循环农业示范工程建设规 划(2017–2020年) (the 2017–2020 Construction Plan for Circular Agriculture Demonstration Programs Combining Grain Production and Livestock Farming). *Gazette of the Ministry of Agriculture of the People's Republic of China, 168*(9), 42–55.

National Bureau of Statistics of PRC. (2001a). 第一次全国人口普查公报(Gazette on First National Population Census of the People's Republic of China). Retrieved November 30, 2021, from http://www.stats.gov.cn/tjsj/tjgb/rkpcgb/qgrkpcgb/200204/t20020404_30316.html

National Bureau of Statistics of PRC. (2001b). 第三次全国人口普查公报(Gazette on Third National Population Census of the People's Republic of China). Retrieved November 30, 2021, from http://www.stats.gov.cn/tjsj/tjgb/rkpcgb/qgrkpcgb/200204/t20020404_30318.html

Plottu, E., & Plottu, B. (2007). The concept of Total Economic Value of environment: A reconsideration within a hierarchical rationality. *Ecological Economics, 61*(1), 52–61. https://doi.org/10.1016/j.ecolecon.2006.09.027

Qi, D., Si, Z., & Scott, S. (2021). Can We Be More Collaborative? Top-Down Policies and Urban–Rural Divides in the Ecological Agriculture Sector in Nanjing, China. *Society and Natural Resources, 34*(2), 208–226. https://doi.org/10.1080/08941920.2020.1774951

Ren, F., & Lin, M. (2016). 生态农业发展的荆棘与路径(Obstacles and Paths in Development of Ecological Agriculture). *Social Sciences Review, 31*(3), 59–64. https://doi.org/10.16745/j.cnki.cn62-1110/c.2016.03.012

Ritchie, J., & Lewis, J. (2003). Qualitative Research Practice: A Guide for Social Science Students and Researchers. In *SAGE*. https://doi.org/10.21511/ppm.14(4-1).2016.04

Shao, Y., Chen, Z., Xiao, H., Zhu, Z., & Li, B. (2019). Integrating Environmental Parameters and Economic Benefits to Analyze the Ecological Agriculture (EA) Application in the Mountain Rice Paddy System of Chongqing, China. *Environmental Sciences Europe, 31*(1). https://doi.org/10.1186/s12302-019-0204-2

Shi, T. (2003). Moving towards Sustainable Development: Rhetoric, Policy and Reality of Ecological Agriculture in China. *International Journal of Sustainable Development and World Ecology, 10*(3), 195–210. https://doi.org/10.1080/13504500309469798

Shi, T. (2004). Operationalizing sustainability: an emerging eco-philosophy in Chinese ecological agriculture. *Journal of Sustainable Agriculture, 24*(4), 113–131. https://doi.org/10.1300/J064v24n04_09

Shi, T., & Gill, R. (2005). Developing Effective Policies for the Sustainable Development of Ecological Agriculture in China: The Case Study of Jinshan County with a Systems Dynamics Model. *Ecological Economics, 53*(2), 223–246. https://doi.org/10.1016/j.ecolecon.2004.08.006

State Council of PRC. (1982). 全国农村工作会议纪要(Minutes of Meeting on National Rural Work). *Gazette of the State Council of the People's Republic of China, 382*(8), 316–326.

State Council of PRC. (1985). 关于进一步活跃农村经济的十项政策(The Ten Policies on Further Revitalizing the Rural Economy). *Gazette of the State Council of the People's Republic of China, 461*(9), 195–200.

State Council of PRC. (1986). 关于一九八六年农村工作的部署(Deployment on 1986 Rural Work). *Gazette of the State Council of the People's Republic of China, 493*(5), 99–106.

State Council of PRC. (2001). 国务院关于印发全国生态环境保护纲要的通知(Circular of the State Council on Issuing the National Ecological Environment Protection Program). *Gazette of the State Council of the People's Republic of China, 1002*(3).

Tan, S. (2019). 推动现代生态农业发展的政策构想(Policy Conception for Promoting the Development of Modern Ecological Agriculture). *Frontiers,* (19), 32–40. Retrieved from 10.16619/j.cnki.rmltxsqy.2019.19.004

UNESCO. (2019). Sustainable development. Retrieved May 23, 2021, from https://en.unesco.org/themes/education-sustainable-development/what-is-esd/sd

Wang, B. (2018). 中国现代农业生态化发展历程与政策导向(On the Ecological Development Course and Policy Orientation of Modern Agriculture in China). *Journal of Fujian Agriculture and Forestry University (Philosophy and Social Sciences), 21*(2), 38–46.

Wang, W. (1998). 中国生态农业的理论与实践(Theory and practice of eco-agriculture in China). *Environment Herald,* (2), 5–8.

Xi, J. (2007). 之江新语 (Zhejiang, China: A New Vision for Development) (1st ed.). Zhejiang People's Publishing House.

Xiao, Z., Zhou, M., & Sun, L. (2005). 中国生态农业模式研究与实证分析(The Theory Research and Demonstration about Chinese Eco-agriculture). *Studies in Science of Science, 23*(2), 208–212.

Ye, Q. (1982). 生态农业(Ecological agriculture). *Problems of Agricultural*

Economics, (11), 3–10.

Ye, Q., & Luo, B. (1987). 生态农业发展的战略问题(A preliminary study on the strategy of the development of eco-agriculture in China). *Journal of Southwest Agricultural University, 9*(1).

Zhang, J., & Luo, S. Himing. (2005). 现阶段中国生态农业可持续发展面临的实践和理论问题探讨(Practical and Theoretical Issues on the Sustainable Development of Chinese Ecological Agriculture). *Chinese Journal of Ecology, 24*(11), 1365–1370.

Zhang, Y., Lin, H., & Li, W. (2015). 生态农业：农村经济可持续发展的重要途径(Ecological Agriculture: an Important Approach to Sustainable Development in Rural Economy). *Rural Economy*, (7).

Zhao, B. (2009). 论道家文化"人与自然"和谐思想的现代启示(On the implications of the harmonious ideology of Taoist "human and nature"). *Journal of Zhoukou Normal University, 26*(1).

Zhao, M. (2019). 中国现代生态农业的理论与实践(The Theory and Practice of Modern Ecological Agriculture in China). *Frontiers*, (10), 24–31.

Zhu, W. (2009). 我国生态农业政策和法律研究(Study on Policy and Law of Ecological Agriculture in China). Northeast Forestry University.

Appendix

Table 23. The interruptions during the interview

Interruptions	Text
Interruption 1	I need to pick up a phone call.
Interruption 2	I have to pick up a call from the City Hall.
Interruption 3	Sorry, I have a call from the mayor.
Interruption 4	You see now the governmental officials from our village come to our farm for a visit. I shall welcome them now and I will continue our interview after I come back.
Interruption 5	Yes, sorry I have so many phone calls. I need to pick up one.
Interruption 6	I am sorry. Here comes again a lot of phone calls (he took a call quickly and didn't hear my question about the harvest)
Interruption 7	(Why do you bother me again? --- Farmer was disturbed by his colleagues) yes, continue

As shown from Table 23, five interruptions were caused by urgent or important phone calls, such as the call from the City Hall and from the mayor. Two interruptions were caused by visitors and the interviewee's colleague. Although most of the interruptions did not cause a significant impact on the interview quality as the interviewer reminded him where they were each time the interview continued, there was still one time when the interviewer failed to express his full sentence about harvest because the interviewee had to take a call, which ended the interviewer's thought on probing more information on this topic. Although remote interview was the best option available during this research, which also created unique challenges, for example the interruptions and quality of internet connection, it is still important to inform the interviewees in the future that use of electronic devices is not allowed until the interview is finished.

Table 24. The analytical framework

Code	Description	Count	% codes
Advantages of CEA product			
AA-level green food	In China, the green food is categorized into two levels, A&AA, AA is the highest level.	1	0.7%
improvement of rice quality	The rice tastes like the ones in the old days when it was grown without heavy fertilizers.	1	0.7%
safe food	The heavy metal test result is significantly lower than the standard of green food.	1	0.7%
rice price	Farmer introduces the rice price from cost to sale.	1	0.7%
rice protein	The protein is stored solid in rice that ensures good quality.	1	0.7%
economic value of crabs	Crabs bring extra income to farmers.	1	0.7%
land problem			
soil compaction	Large amount of soil organic matter is decomposed.	2	1.4%
increase in use of fertilizer	More fertilizer is used to achieve the same amount of production.	3	2.1%
reason for engaging in CEA	Farmers found the land couldn't produce crops.	1	0.7%
land lost fertility	Heavy input of fertilizer leads to loss of fertility and crop production.	2	1.4%
concern from the government	The Chinese President is worried about the land problems with heavy use of chemical fertilizers.	1	0.7%
pollution from chemical fertilizer	Chemical fertilizer pollutes the water.	3	2.1%
Technologies used in CEA			
legacy from ancestors and use of biotechnology	Green manure and straw-returning to the field are methods from ancestors and biotechnologies are used to protect plants.	1	0.7%
technique for cultivation	adding plant-based materials, a type of modern technology, to increase the minerals need for soil fertility.	1	0.7%

Table 24 cont. The analytical framework

Code	Description	Count	% codes
example of increasing microorganisms	Farmer introduces one technology used for the field.	1	0.7%
The use of biotechnologies for pest control	Use of plant-based pesticide to kill pests.	1	0.7%
rice-duck farming system	Rear of ducks in parts of the field to kill insects that serve as natural protein source. They are reared for sale too.	1	0.7%
role of fish in CEA	Fish feces provide nutrients to the crop.	1	0.7%
rice cultivar	selection of traditional cultivar that is non-GMO, rich in nutrients and tastes sweeter.	3	2.1%
crab rearing	crab rearing as part of the rice-fish-crab farming system.	1	0.7%
conservation during fallow period	Use of green manure for soil recovery when the land is not used for growing crops.	1	0.7%
Technical explanation			
altitude and temperature difference	altitude and temperature difference in the geographical location of the study	1	0.7%
microorganisms	To increase the amount of microorganisms is important for soil fertility.	1	0.7%
expert's knowledge in choosing fertilizers	The agricultural experts recommended a series of fertilizer for use in the field.	1	0.7%
long plant growth cycle	The long growing cycle for rice on the farm under natural conditions.	1	0.7%
technical explanation of current CEA	Farmer talks about the nutrients of the soil and two circular modes of farming.	2	1.4%
Farmer's perspective of CEA			
adaptation to nature	Human shall adapt to nature and follow the laws of nature.	2	1.4%
first step in CEA implementation	producing good quality rice.	1	0.7%

Table 24 cont. The analytical framework

Code	Description	Count	% codes
second step in CEA implementation	Improving the values the land can provide.	1	0.7%
non-chemical farming	A form of Chinese ecological agriculture that prohibits use of chemical fertilizer and pesticide.	1	0.7%
soil has life	Soil has life that is not visible with naked eyes.	1	0.7%
recognition of the harmony between man and nature	Sustainable agriculture shall follow the laws of nature and man shall adapt to it.	1	0.7%
Taoist's view of human and nature	Human shall follow the laws of nature and adapt to it.	1	0.7%
care for the next generation	Providing healthy food for the next generations.	1	0.7%
contribution to Rural Revitalization	The ecological farming helps revitalize the land and make farmers better off.	1	0.7%
farmer's conclusion	1. good seeds and good technology cultivates a good field. 2. farming should pursue the harmony between man and nature.	3	2.1%
Responsibility			
being responsible	Farmer takes the responsibility and covers the loss.	1	0.7%
commitment to compensate loss	Farmer made a commitment to cover the financial loss of other farmers if their crops cannot be sold at the agreed price.	2	1.4%
cover the cost and honor the commitment	The crops didn't grow as planned and the farmer had to compensate the financial loss of other farmers as promised.	1	0.7%
harvest guarantee	The food production shall not decrease.	1	0.7%
Other farmers			
concerns from other farmers	Other farmers care about the technology for use in the farming and the economic profit of rice.	1	0.7%

Table 24 cont. The analytical framework

Code	Description	Count	% codes
economic lure to persuade other farmers	The farmer persuaded other farmers to join him for the new type of farming by giving them promises of more income.	1	0.7%
other farmers' feelings of CEA implementation	They feel it not only brings them healthy food and more income but also brings the ecosystem back to life.	2	1.4%
Problem in CEA			
loss of crabs due to distant transportation	Almost half of the crabs died when they reached the farm.	1	0.7%
insect hazards	Insects appear in the field when the land recovers its fertility.	1	0.7%
insufficient fertilizer leads to problems	The experiment showed that the non-chemical fertilizer was not sufficiently applied and crops didn't grow.	1	0.7%
Farmer's interaction with other farmers			
farmer's reflection	1. future generations shall have healthy food to eat 2. more income for farmers motivates them to believe and engage in ecological agriculture.	2	1.4%
help from experts	Agricultural experts help train the farmers and motivate them.	1	0.7%
no voluntary participation	Other farmers didn't want to participate in the beginning.	1	0.7%
sharing dividends	Fair sharing of profits among farmers.	1	0.7%
trust of other farmers	The farmer wins trust from other farmers.	2	1.4%
happiness of harvest	Farmers are happy to see healthy crops ready for harvest and for the business sale, which brings them income.	1	0.7%
Farmer's interaction with the interviewer			
farmer's invitation	Farmer invites the interviewer for a visit to his farm.	1	0.7%
farmer's honesty	Farmer is honest with sharing things with the interviewer.	1	0.7%

Table 24 cont. The analytical framework

Code	Description	count	% codes
farmer's generosity	Farmer gave rice to his visitors.	1	0.7%
farmer's gratitude	Farmer is thankful for the interview.	1	0.7%
farmer sharing a lesson	Theories must be verified through practice.	1	0.7%
distracted farmer	At this moment he did not show up in the screen as he was looking for something.	1	0.7%
reluctance to recommend other farmers for interview	1. other farmers are unlikely to express themselves clearly. 2. farmers are busy with agricultural production.	1	0.7%
Pioneer			
being a pioneer	be the first one to start the non-chemical farming.	1	0.7%
courage to try something new	The farmer is brave to try the farming technology recommended by the agricultural experts.	1	0.7%
determination to implement CEA	After a few years of practice, the farmer decided to implement the non-chemical farming regardless of support from the government.	1	0.7%
farmer's gratefulness	Farmer is grateful to his supporters.	1	0.7%
experienced farmer	The farmer has almost 30 years of farming experience.	1	0.7%
Farmer's agenda			
calling for sustainable change in agriculture	The situation of Chinese agriculture with heavy chemical fertilizers input should stop and be replaced by the one with non-chemical fertilizer for the health of the future generations.	1	0.7%
healthy food for future generation and economy	Farmer aims to produce healthy food and market it for better economy.	1	0.7%

Table 24 cont. The analytical framework

Code	Description	Count	% codes
merits of CEA	CEA benefits the current and future generations.	1	0.7%
transition of CEA	The non-chemical farming aims to align with governmental policies and adapt to nature.	2	1.4%
rice pilot project	collaboration with provincial government to grow rice.	1	0.7%
Farmer's love			
dedication to agriculture	The farmer puts all of his time into agriculture and agriculture is part of his life.	2	1.4%
love for the land	The farmer has a deep feeling for the land and dedicates his life to it.	4	2.8%
meaning and value	Doing something meaningful and contributing to the society.	1	0.7%
Changes after CEA implementation			
recovery of vitality and life	When the soil gains fertility, the crop grows healthily and the arable layer gets deeper.	1	0.7%
recovery of ecosystem as incentive	The recovery of ecosystem due to improved soil fertility makes the farmer believe in CEA.	1	0.7%
increase in production	The non-chemical farming system produces more crops per area than conventional farming.	1	0.7%
separate codes			
subsidy for CEA	National subsidy that encourages use of bio-organic fertilizer.	8	5.7%
exploitative production	increasing the use of fertilizer to create more food at the cost of the environment and the soil.	1	0.7%

English translation of transcript

Interviewer: A

Interviewee: B

A: Hi, I'd like to first introduce to you the theme of this research topic. It is about values in Chinese ecological agriculture. When I developed this topic, I was communicating with my German supervisor about how values can play a role in ecological agriculture, specifically in the context of China. Therefore, as I was searching online, I found a report about you in Liangshan news (a local news media) where you were interviewed regarding the rice-fish-crab farming system. To my surprise, I did not realize that there is already ecological agriculture in Xichang. I was very happy and wanted to contact you. However, I am in Germany and am not able to call you with my German phone number. I turned to my father and asked him to find you, with the information I found online. I think it is meaningful to communicate such stuff to the wider public, as China is on the path towards sustainable development and construction of ecological agriculture. Therefore, it is also meaningful for the contribution to rural areas and farmers. Okay, for this interview, if you are unable to answer some questions or you need time to think about it, feel free to take the time and also answer questions only willingly. You can tell me directly and ask for clarifications in case I don't make myself understood, okay? Shall we begin?

B: Let's begin.

A: Okay. Well, first I'd like to know, as I am not so familiar with the geographical and historical conditions of You-jun village, thus, I have some initial questions about your farming cooperative. When was the cooperative established and how many farmers have joined it?

B:(pause), the cooperative was established on 6th of August, 2015.

A: Okay.

B: There are 403 households in the cooperative.

A: Did you organize and set up the cooperative? Or, how have you made them join it?

B: Hmm, it was… let me first introduce under what conditions the cooperative was set up.

A: Yes, sure.

B: I am a typical farmer and I have almost 30 years of farming experience. Before 2015, in the period between 2006 and 2015.

A: Yeah (nodding).

B: We, well, I firstly found that the soil in our land, why when applying the fertilizer, we needed to increase the amount of fertilizer each year in order to achieve the same amount of production, however, well, we found one problem, oh I need to pick up a phone call

A: Okay. Go ahead.

B: Hello!

A: Hello!

B: I am so sorry, these days there are a lot of phone calls because of the construction (of farm). Before the interview, I finished my meeting and came back to the factory … okay let's continue.

B: From 2007–2015, I came to realize that we needed to increase the use of fertilizer each year in order to achieve the same amount of production. This led to one, no two consequences. First, after use of large amount of chemical fertilizer, our land lost its fertility, with a reduction in its soil organic matter (SOM). In 2016, actually in September of 2015, we tested the soil and found that the SOM was at 1.1%. My teachers from the lab found that the main nutrient of the chemical fertilizer is nitrogen, which consists 46.6%. When nitrogen goes into the soil, it mainly decomposes the SOM etc, along with other micro elements, to provide nutrients. In the end, what happens? Large amounts of SOM are decomposed, which leads to soil compaction. The microorganisms decreased sharply. The land, as well as the soil, has life, although it cannot speak and you are unable to see it with naked eyes. Yes, however, it will only answer you with the crops you have sown—it lacks nutrition. In 2016, we started the non-chemical farming system, back then my teachers proposed a new term "organic stuff comes from non-chemical farming." What does "non-chemical" mean? It means prohibiting the use of chemical fertilizer and pesticides in the crop farming system. After few years of such practice, the SOM in our field has reach 2.1%, 1% higher than before, a significant increase.

A: Yes(nodding)

B: This means our soil has over 200 million of active bacteria per gram, which is great. In this non-chemical farming, we continue to wisely use the legacy from our ancestors, such as the green manure and straw-returning to the field. We have planted hairy vetch as green manure which serves as the rice fertilizer for next year. In addition, we applied the so-called "placeholder bacteria for plants," which is a type of useful bacteria that occupy the surface of the plant so that other bacteria, particularly the bad ones, are unable to reach the plant. Thus, the plant is able to grow better and such bacteria can also prevent soil diseases for the second year. Since the soil is fertilised and crops grow healthily, the land recovers its vitality and life. How do you know it? First, the arable layer gets deeper.

As the plants grow up, they appear strong and not sick. The improvement in soil fertility helps recover the ecosystem. The animals and various fishes which disappeared years ago now come back, therefore, it is a good sign. I believe in this type of farming and I also communicated this with my fellows when I went to different conferences for exchanges.

A: Yes.

B: It is important for me to protect the land that produces healthy food for our future generations, and I also believe that to market the healthy food grown from the land will contribute to our economy as well as ourselves. How can the citizens become strong? First they need to have a strong body. It comes from the food. When I was attending a conference in Shanghai, I expressed that the future of agriculture in China cannot continue with the reliance on heavy chemical fertilizers. Otherwise, in 30 years we all become sick man of Asia, and this time it is without any invasion from other countries.

A: (nodding)

B: As regards the increase in production, Mr Wu, I have to tell you that if it is not because you are also from Xichang, I will not tell you the secret at all.

A: Oh thank you!

B: Therefore, compared to the conventional farming, the one that uses chemicals, our farming system is able to increase 100kg (per mu/Chinese unit of area, equivalent of 0.165 acres). This is a significant number for field-crop production. The advantage also shows in the quality of the rice. One can feel the taste of rice grown back in 1970s and 1980s (when heavy fertilizers were not in use), the smell of authentic rice. As for the diseases, in general there are hardly diseases, except for the insect hazards, which are annoying. Insects are also organisms and when the environment gets better it improves the biodiversity. In our field there are loaches and eels, therefore there are insects too. The insects like to eat the green leaves in the field because they are tasty. This is the only hazard that bothers me, however, such insects are not drug-resistant.

A: (nodding)

B: Take the rice borer and armyworm for example, we use azadirachtin, a type of plant-based pesticide, which is extracted from neem seeds. The pesticide contains 3% of azadirachtin. After applying it to the plants, the insect that consumes the leaves will absorb it in its stomach.

A: (nodding)

B: For the first two days there are no noticeable changes of the insect, however, after 3 to 7 days the insect will gradually stop eating, meanwhile it becomes to shrink and lose its appetite, on day 10 the insects will fall dead. The most important fact is that we use plant-based pesticide that does minimal harm to the plants, so this is quite good. The good thing about this reflects from the test result that qualifies our rice as AA-level green food (in China the green food is categorized into two levels, A&AA, AA is the highest level). We know that the organic certification comes from the EU, and the AA-level green food certification is the result of organic certification combined with Chinese conditions. The organic farming applies strict regulations in each process; however, the non-chemical farming system uses plant-based and bio-based materials according to the local conditions. Thus, after a few years of such farming practice, we came to the conclusion that the future of Chinese agriculture should promote it, regardless of support from the government. The non-chemical farming is initiated by us, the very first one in China. When I visited some farms in Chongming island in Shanghai in 2018, I found that the price of rice that is produced without use of chemical fertilizer and pesticide is worth almost 200 RMB per kg.

A: Yes.

B: Our conclusion is that good field and good seeds are necessary. Good seeds, good techniques and good field—these three *good* are necessary.

A: (nodding)

B: In the following years, well, first, the farmers from our cooperative feel they eat healthy food. Our country has a huge population and consumes a lot, and since food feeds our people, we have been continuing to implement the non-chemical farming system. Mr. Wu, here comes some critical aspects. First is the carbon element that many experts and scientists ignore. For carbon, we know that there are carbon, hydrogen and oxygen in the air, as well as nutrients like nitrogen, phosphorous and potassium. And we know that dews that are brought by the air to the crops can help them grow better. This is because the crops can absorb such nutrients through photosynthesis, however, the nutrients from the air are not enough, so we have to supply some using raw materials based on biomass. We use bio-organic fertilizer that meets the national standard NY884. Remember that this type of fertilizer is used in our field. It comes with an expensive cost, around 4000RMB per ton. We made another conclusion, that is, we shall promote the type of agriculture that brings harmony between man and nature. Man takes his law from the Earth; the Earth takes its law from Heaven; Heaven takes its law from the Tao. The law of the Tao is its being what it is. It all boils down to nature but we cannot simply do nothing. (The point is), we do not interfere the laws of nature and we shall adapt to it. According to Laozi (the founder of Taoism), the law of the Tao is its being what it is.

The harmony between man and nature, as our great leader President Xi puts it, the agriculture that takes care of such harmony is to be the sustainable agriculture for the future of China. Therefore, when he proposed the concept of *community of shared future for mankind*, it encompasses all sectors including agriculture. As we started this farming practicing since 2016, we are on the road of exploration. This year our cooperative will collaborate with the Sichuan-provincial government for a rice agriculture pilot project. This is good, because our future generations can therefore eat healthy food. Remember the time when chemical agriculture came into China in 1930s. Back then the chemical fertilizers are called "powdered fertilizer," with carbon and nitrogen being its main substance.

A: (nodding)

B: Oh, since the economic opening up in 1980s, let me tell you Mr. Wu, nitrogen fertilizer was applied heavily in the field. If my memory doesn't betray me, from 2001 we applied 40kg of nitrogen fertilizer per mu, later we added it up to 60kg. In this way the SOM was largely dissolved and caused soil compaction. The food production was really not good.

A: Did you only grow rice? Was it monoculture?

B: Nope. It's a circular mode now. It means we use rice, fish and crab as the subject of such farming system. We have another feature here in some parts of the field where we grow rice and duck together. For each mu we grow 8 to 12 ducks which we purchase from the market. As the ecosystem recovers, there are a lot of freshwater snails too. If you come back to China this year and still consider the chance to exchange ideas with me, then you are welcome to come to our field and see it yourself.

A: Oh yeah of course!

B: Our freshwater snails, actually the ducks, they are reared in a free-range way. Most of the ducks on the market are sold for only 8 to 10 RMB per 0.5kg. Our ducks are sold at 200RMB per head. We put baby ducks in the field in May and they are all sold out by winter solstice. It is not possible to save them for later sale in Chinese New Year. This means it is very good because as I told you the stuff about ducks. Since our ecosystems recovers the biodiversity also improves, a lot of armyworms appear in our field and they like to feed on the leaves of rice. We rear ducks as a strategy to eat the armyworms which serve as good protein source. The ducks eat them so much that they can barely walk.

A: (laughing)

B: The duck meat is rich in amino acids, particularly the essential ones, and the amino acids decide the taste so we call it flavory amino acids. The duck meat is very delicious and for this reason it has a high price. As for the rice in our field, the cost of the rice is 3 RMB per 0.5kg, after processing and packaging the price rises up to around 8 RMB per 0.5kg. The wholesale price in our region is 10 RMB per 0.5kg. We use the type of rice that adapts to the local conditions, the one that is traditional and is not GMO.

A: (nodding)

B: The day when your dad visited me, I gave him a few packs of rice as gift. I told him that the pleasant smell of the boiling rice can reach area of 20 to 30 square meters!

A: (laughing)

B: Yes, and yeah, I thank you for offering this opportunity to let me introduce this type of CEA. An important aspect to consider is the altitude. Here in Xichang the altitude is 1,425 meters and the temperature difference between day and night is significant, which results in high quality of dry matter accumulated from photosynthesis. When the carbon supply for photosynthesis is enough, the formation of carbohydrates will be tighter, as shown in the presence of crystal protein.

A: Okay, I …

B: Also from a micro level, the nutrients are aligned in molecular forms and are in order, not in random form. It is therefore stored in solid form in rice. The quality is guaranteed. As for the rice and crab system, for each mu, we rear 400 baby crabs, each weighting 23 g on average. The crabs mainly feed on the freshwater snails. We start rearing crabs from May to July, according to the natural conditions, the laws of nature. And we harvest them at the end of October when they grow big enough, around 155 to 130g, and on average each crab weights 118 g. For the economic value of such crabs, as we say from a local citizen's view, since the crabs have more bones than flesh so there's less to consume, however, with the higher demand on market and as our living quality improves, we can sell them at 37 RMB per 0.5kg for wholesale price. For each mu, these crabs bring several thousand RMB to farmers. For rice-crab farming system, since we purchase the baby crabs from Zhejiang province, when they reach here quite some crabs are already dead, actually to the extent of almost 45-50% of the total purchase, and this is the situation when the transportation process was conditioned by cooling ice. When the crabs reach our farm, it will take a couple of days before putting them into the rice field. This year we experienced a huge loss of crabs due to the long-distance transport. However, such farming system is already recognised and from next year onwards we will promote the rice-duck-crab farming system, which is also a circular and ecological farming system.

A: I have a question. Are the baby fishes also purchased from Zhejiang province, or you use the local baby fishes?

B: No, we use the local ones, the common carp.

A: Okay.

B: Common carps need to eat to survive. Since it eats, it will evacuate faeces that provide natural nutrients to the rice crop. The same logic applies for the rice-duck farming system.

A: Yes, in this way the energy and nutrient flows are circulated. I have another question; do you plant rice once a year or several times a year?

B: Once per year.

A: Once per year.

B: Once per year.

A: Okay.

B: We plant rice once per year because the plant growth cycle in this region is rather long. Different from other regions such as in Zhejiang Province where it takes 120-130 days from planting to harvest, and Chengdu plateau where it takes 140-150 days, here in our region it can last up to 197 days.

A: I see.

B: But it also depends on the cultivars. Our rice type has enough dry matter so it is rich in nutrients.

A: (nodding)

B: The taste is sweeter, softer and chewy.

A: (nodding), I asked this question because I saw news about the rice-fish farming system where the farmers from somewhere I forgot, he planted the rice two or three times per year in order to increase the economic profits, so I…

B: I am sorry Mr. Wu, I have to pick up a call from the City Hall.

A: Alright.

B: Yes, I am sorry.

A: It's okay.

B: I am sorry, I had a call from the City Hall and they wanted to come here this afternoon for a survey.

A: No, it's okay. I understand that farmers are busy. Well, I was just going to talk about that some farmers who engage in rice-fish farming system plant rice two to three times a year, so I asked that question.

B: Here we grow rice once per year.

A: Yes, I see.

B: If you want to ask more questions, then hurry up.

A: Yeah, I do have several questions. I wonder if you can describe the situation of harvesting the fish and crabs. Is this activity only involving farmers, or are citizens also invited to participate?

B: Well, hmm, think this way. According to different varieties of crops we harvest in different regions. The situation of harvest is simply full of happiness, we are happy to see the crops are ripe and without any diseases. It's a happy feeling. The farmers are smiling during the harvest time because the food production is increased and their money pocket will be filled. Plus, we plant crops according to how many business orders we have.

A: Right, I see. Another question is about the cooperative. How were you able to gather the farmers together to engage in this type of ecological agriculture? Did they look for you or you persuaded them?

B: Well, this is a critical question. At that time, they knew that since my ancestors all come from this region so I am trustworthy. The first reason was because we are all farmers and the land has not been able to produce crops for a long time. I told them this situation and contacted the agricultural experts from Xichang University. We had a meeting here to train and motivate the farmers to learn how to engage in such farming system. And I also promised them that if their crops are not able to achieve the target production amount, then I will cover their financial loss.

A: (nodding)

B: The second concern is the choice of fertilizer. The experts from the university recommended a series of fertilizers to choose from. For the first year, we chose several brands. But out of the several brands we prioritized one for use in the field because that one stands out with its nutrients and indications. In the end we chose five fertilizers in total and one of them is used for our trial field. In the first year the field size is 836 mu. We hoped this would work and can promote it in the later stage.

A: I see.

B: For the first year, we used …, Mr. Wu, remember what I tell you, doing research is a critical work. Only with practice can the research draw a conclusion. Keep this in mind, this will help you avoid entering a wrong direction, no matter in which academic field you will be in the future. In the first year when we held our meeting with the farmers, the experts told them that according to the relevant data provided by the fertilizer company, each mu needed only 120Kg of fertilizers for the crops to grow well. Then, okay we followed the experts' instruction and applied 120kg of fertilizers. Then we found the nutrients were not enough because the rice did not grow. The other farmers all came to me for explanation. Imagine it is a big field of 800mu! I had no choice and told them to apply one more pack of fertilizer for each mu. And guess what? We had a good harvest that year. I remember the price of fertilizer was over 100RMB each pack and I personally spent one to two hundred thousand of RMB to cover the cost.

A: Right.

B: Since it was a harvest and we found the right method, in the second year we increased our pilot field to 1953 mu. We increase the field according to the number of business orders we receive.

A: Which year was it when you said that the expert told you …

B: The first year.

A: Okay.

B: The first year. The experts, well, they calculated the amount of fertilizer for use according to the nutrients content. Sorry, I have a call from the mayor.

A: Okay (frustration).

B: You can continue with questions.

A: Did you mean the year 2016 when you mentioned the first year?

B: Yes, yes, yes, 2016.

A: Right okay, well …

B: Continue talking, I hear you. (At this moment he did not show up in the screen as he was looking for something.)

A: Yes, about the non-chemical farming. For the use of such technology, I wonder if the national or local government provided any subsidies or policy incentives?

B: Well, there are significant incentives from both policy and project support. Specifically, for use of bio-organic fertilizer, each mu is subsidized by dozens of RMB.

A: I see.

B: The bio-organic fertilizer we are using for the field costs at least 600RMB per pack, and we get more than 100 RMB for the subsidy.

A: Then do you think this subsidy …

B: Including policy incentives.

A: Please go ahead.

B: This subsidy comes from relevant policies and projects that tilt support towards use of such technology in agriculture. It encourages you to use such fertilizers. Actually, each Chinese President knows there is something wrong with our land and soil.

A: (nodding)

B: Yes, that's why when President Xi went to Northeast China last year, he mentioned the need to protect the black soil. He went to Heilongjiang province and made a speech there. None of our former presidents has ever mentioned the protection of the black soil. But President Xi mentioned it.

A: Right. Back to what you said earlier, since the rice is grown once a year, how do you protect the land when it is in the fallow period?

B: For some areas we flip the surface soil over so that it can absorb the nutrients from the air, for other areas we plant some hairy vetch as green manure.

A: I see. In this way the soil is protected, and …

B: Exactly, the fallow period is used for soil protection and recovery.

A: (nodding), and it also serves as the preparation for the next year plantings.

B: Exactly.

A: Also, I wanted to ask a relevant concept. You might know that President Xi also proposed the concept of "clear waters and green mountains are as valuable as mountains of gold and silver" in 2005. At that time, he visited a village and mentioned that the rural development of that village was worrying because the environmental quality was terrible; later the villagers decided to keep the clear waters and green mountains so that they can get mountains of gold and silver. I think this process is similar to what you are doing to the field. After all you firstly adopted the method of non-chemical farming, in which the pollution is reduced as much as possible, and this farming system is also friendly to the soil. Actually, this type of farming is already moving towards the sustainable transition. As you mentioned the fallow period, we know that the field is not used to produce any economic profit during this period, whereas it is conserved so that it can continuously support ecological farming in the long run. I think this is of great importance, that's why I asked you if the rice is planted once a year instead of two to three times a year. I also agree that we cannot increase the use of fertilizers in order to create more economic benefits.

B: Yes, you are talking about the exploitative production, which in my opinion is wrong and will not work in the long run.

A: Yes, right.

B: Yes, it is also wrong to simply do nothing. We shall adapt to nature and follow the laws of nature. The best comes from what grows in a natural way.

A: Yes, it follows the laws of nature.

B: Exactly, follow the circular laws of the natural world, that's right.

A: Yes, I think you mentioned one important aspect—that the land has its own life. Since human are part of the nature, right?

We learn how to stay in harmony with it, which is very important. I am also surprised that compared to other farmers whose education level is limited and who might have difficulties expressing ideas, you seem to have no such problems. I wonder how many farmers like you are there in your cooperative, I mean those who both know the technology and know how to put it into practice.

B: First let me answer you with this. The main thing is that one has to feel attached to the land, love the land, and proactively learn from experience. Yes, and in this process one also combines with his own practice. In my case, I also got indispensable help from the experts.

A: (nodding)

B: I describe myself as a farmer who deeply loves the land, and I also see it in the long run. I think of the development in the next 10, 20, 30, 50 years. After the efforts from our generation, the next generations can enjoy a life with healthy food. This is what I consider.

A: (nodding)

B: This year I am turning 50 and I cannot consume a lot of food. For me the main thing is about the meaning and value as an individual. We come to the world with nothing on our hands, but we want to contribute something meaningful to the society before we pass away. For agriculture sector, there must be someone who starts doing this, and I would like to start this. In every sector there are people who understand policy and are equipped with practical skills. There are also farmers like us. However, it depends on the extent to which the individuals accept the concept.

A: (nodding)

B: As regards farmers' education level, it is hard for me to comment. Yeah, you know, my teachers are excellent, including the experts and principal from China Agricultural University. The principal visited our farm and he asked if I am a farmer, I said yes I am. He did not believe it and thought I was someone from the City Hall who was designated to explain to him the situation of the farm. It was until later when he went to my house for meal and I showed him my ID card, that he believed I am a farmer.

A: (laughing)

B: This is true!

A: I believe you!

B: Now I can clearly tell you that China is actively training farmers who can understand policy and are equipped with practical skills. However, it is up to the trainees' own attitude and perception that can decide the result of such training programs.

A: Right.

B: For me I can simply say that I love agriculture and I love the land. People call me crazy because I don't drive my car, instead I walk around the farm to check the conditions.

A: So, how did you come to know the non-chemical farming practice and the rice-fish-crab system? Was it out of interest and you looked for it somewhere?

B: I just felt the need to plant the rice in a good way.

A: (nodding).

B: With a stroke of luck, I came to know my teachers from Xichang University. They introduced this technology and asked me if I dared to give it a try. They also mentioned that it must be implemented to see if it would work well. I told them yes, I would try. However, it has not been always easy and straightforward. Like I told you earlier, we used four packs of fertilizer instead of three. In the implementation stage we managed to produce good quality rice, this is the first step. The second step is to improve the values the land can provide, since the land is precious and its resources are non-renewable. Then we contacted the dean of animal science department of Xichang University. From there we know the rice-fish-crab farming system. We also grow rice and ducks together since there is already good reputation about the Jianchang duck. The experts from the animal science department provided technical guidance on the insects and pest control, and with this support we are able to take good care of this farming system. Thus, we also contribute to the Rural Revitalization. Rural Revitalization, which we are part of, is to revitalize ecology, build up rural civilisation, cultivate an environment where people are honest and friendly, and make farmers better off. These are the essential aspects of Rural Revitalization.

A: And when you came to know …, as we talked about the subsidy from the national government, I wonder if you ever thought about subsidy before you started rice-fish farming system, or you came to know the subsidy after you started it and you felt more motivated to continue doing it?

B: I knew the subsidy after I started this farming system. Mr. Wu, you see now the governmental officials from our village come to our farm for a visit. I shall welcome them now and I will continue our interview after I come back.

A: Alright.

B: Okay.

B: They are late and will come in half an hour, how about we continue with the interview?

Okay, going back to the question, I was asking if you received the subsidy after you started the non-chemical farming system, or you knew the subsidy first and then decided to do this?

B: No, I have already done it for two years before there was such subsidy.

A: So it means you applied for it later, or they issued you the subsidy when they came to know you were doing such farming?

B: Yes (to the first one).

A: Okay, so you applied for it. I see. Do you know if there are specific policy documents that support the non-chemical rice farming? Are there more local policies or national ones?

B: So this subsidy comes from a national-level project that indicated the need to decrease the use of chemical pesticides and fertilizer, and to increase the use of bio-organic fertilizer.

A: I see. So it is a specific project targeting chemical pesticide and fertilizer.

B: Yes, to reduce the use of chemical fertilizer and pesticide, and increase the use of bio-organic fertilizer.

A: (nodding)

B: It means to decrease the use of chemical pesticide and fertilizer while improving the soil fertility by applying bio-organic fertilizer.

A: (nodding)

B: Under such conditions, it also has to guarantee that the harvest is not compromised and farmers' revenue is increased.

A: For this policy document, do you have it or do you know where one can find it? Perhaps from online?

B: Well, you can search it online using keywords such as the decrease in use of chemical fertilizer and pesticide, it's easy to find it.

A: Ok, and it is a national rather than local policy, right?

B: Yes, it is a national policy.

A: Ok, this is good. Hmmm, let me think, I have another question: how did you persuade the other farmers to join the cooperative? Did they volunteer to participate when they heard this project? Or after your persuasion ...

B: No one volunteered to participate.

A: No one.

B: The main idea was, you know, in that year the market price of rice was only 1.25 RMB per 0.5kg. I told the other farmers that if they participated in this project the rice they grow would be sold at 2 RMB per 0.5kg. I gave them this promise and under the lure of economic benefits ...; I also told them if the rice cannot be sold with this price, then I would personally cover their financial loss.

A: In this way they joined the cooperative.

B: I have to shoulder the responsibility, since I am the initiator.

A: Right, I see. This sounds quite challenging. So was it in 2016 or before?

B: From 2015 I started to talk to the farmers about my idea.

A: I see.

B: The first thing is that they have to trust you. I won their trust because of my interpersonal relationships and my moral qualities as a person so they are assured they can contribute part of their land to this project. 800 mu is a big number!

A: Exactly.

B: Imagine in the worst case, if there is something wrong, the other farmers have to consider if I am able to cover their loss as promised.

A: Yes, yeah, did the other farmers grow only rice in their own field? What did they grow?

B: Only rice, the hybrid rice.

A: I see.

B: The common ones that you can get from the market.

A: Right.

B: Here in our field, we use good seeds and good technology to cultivate a good field.

A: (nodding), so it means that the other farmers contribute part of their own land to your cooperative for implementing such farming practice?

B: Exactly, we share the dividends based on how much rice one contributes to the cooperative. I charge nothing from their profit. After a few years of dedication, now there are a large group of people supporting me, and I am sincerely grateful to them.

A: I see. How did they feel after they joined you? I mean, except for the economic profit, did they feel that engaging in non-chemical rice agriculture meant something more than just income growth? Perhaps they also feel attached to the land? Or they simply feel the main thing is to earn more money?

B: Well, the economic benefit is the main reason. Otherwise, they feel that the ecosystem is improving, for example the water in the ditches is becoming clear.

A: (Nodding)

B: Yes, each year I hold training sessions with them where I ask them this question: have you ever thought about the fact that since our generation gradually enters old age, what can we provide to feed our next generation? Are we assured that what they will eat will be healthy? This is a question. What concerns them the most is the leadership of the cooperative. First, they do not have to worry about the technology. Second, the rice they grow will sell for a good price. According to my report to Hu Chunhua, Vice Premier of the State Council, this type of farming system advocated by farmers is a project that not only benefits current people's livelihood but also takes care of the future generations. It merits priority and support from the government.

A: Okay, so before the non-chemical farming was implemented, the environment, water ditches and water quality, as you mentioned earlier, might be influenced by the monoculture. Do you remember what kind of environmental problems were there before the ecosystems were recovered? Any specific aspect?

B: Yes, sorry I have so many phone calls. I need to pick up one.

A: Oh, okay, go ahead.

B: (When the interview resumed, we didn't say hello but continued with the last conversation) well, believe this. When large amounts of chemical fertilizer are applied in the field, the soil can only absorb as less as 13% of it.

A: (nodding)

B: For the rest of it, at least 50% evaporates in the air in the form of molecules, such as CO_2. Therefore, more than 10% stays in soil, another 10% is taken away by water, and for that part it will pollute the water.

A: Yes, the water quality was not so good.

B: No, it was not. But now it is good.

A: Oh, was there no air pollution etc.?

B: Of course there was; you have to believe this.

A: (nodding)

B: The CO2 indicator from the air must be significantly higher, but human cannot perceive it. It's not that you inhale the air then you get poisoned. It is not poison.

A: (nodding)

B: The fertilizer goes into the air, water and soil, but the majority of it is washed away by the surface water.

A: (nodding)

B: When the water runs away, it absorbs large amount of the fertilizer residues remained in the soil. Mr. Wu, in simple terms, the nutrients go back to the natural world.

A: Yes.

B: Right.

A: As regards the harvest, you said ...

B: I am sorry. Here comes again a lot of phone calls (he took a call quickly and didn't hear my question about the harvest). Therefore, no matter what it is, even if it is human, they start to decompose the day they die, and the decomposition process goes back to the natural world. Some parts of the process are not visible to the naked eye. Other parts get evaporated when exposed to the sun or blown away by the wind. They become eventually nutrients.

A: (nodding)

B: It returns to the natural world.

A: I have another question. In your daily life, except for work where you go with other farmers to the field, are there any other moments when you feel like going there for a walk, because of the beautiful view etc.?

B: (Why do you bother me again?—Farmer was disturbed by his colleagues) Yes, continue.

A: Yeah, I just asked if you can describe what is your routine when you work, and when you will go to the field except for work. Do you go for a walk to relax yourself?

B: Ok, I understand this question. I put it in a simple way. I am in the field almost at any time because I love agriculture and I feel like going there. For me there is no difference between work and leisure time. When I think about the plants in midnight, I go there. I want to see how the plants change during 11 pm to 1am and 11am to 1 pm. I check and record the growth of rice ears, temperature etc., and I summarize my own learning, so that I can share it with other farmers when they need guidance.

A: Right.

B: Therefore, for me there is no such thing as work. I am always there, day and night…

A: In the field.

B: Yes, in the field.

A: Right, I see. Hmm, do you have further plans regarding the non-chemical rice? Have you thought about expanding it?

B: My next step is to make it well-known how the Chinese ecological agriculture shall develop. The farming system we are doing here is transitioning towards a model that aligns with governmental policies and adapts to the nature.

A: (nodding)

B: In terms of expansion, I wish this model can be implemented everywhere in China. The obstacle is that it is still a tough and long road to go. This is my answer.

A: Right, after all, to grow crops in such a way requires first a good field. The soil must be good enough. I wonder what kind of techniques you have used for conservation of the field?

B: For techniques, we know that good field and good soil needs cultivation. How do we cultivate it? We do it with the help of modern technologies while respecting the harmony between man and nature. The cultivation method is traditional and obeys the laws of nature. When some minerals are not enough, we add plant-based materials instead of chemicals to the soil. A good example is the hairy vetch. I mentioned this earlier.

A: (nodding)

B: Yeah, improving the proportion of carbon and nitrogen so that the soil becomes porous and soft. The most important aspect is to increase the amount of microorganisms.

A: (nodding)

B: When the soil becomes fertile, good crops grow out of it. When there are good crops, we can provide good quality food on the market, and ultimately the farmers get more income which makes them believe in such ecological agriculture that obeys the laws of nature and takes care of the harmony between man and nature.

A: Yes.

B: Mr. Wu, let me introduce you to a very simple method we use in our field. You know the faeces of cow right? Here we have faeces of cows and sheep. There are nutrients inside but the variety of nutrients are quite few and there are also insect eggs and worms in it. What we do is to apply human interference by adding the microorganisms and fermenting the mixture in high temperatures between 73 to 85 degrees Celsius. The fermentation takes three days and the microorganisms multiply. Then we apply the manure to the field where the microorganisms become more active. The key part is to provide enough nutrients to the soil because the microorganisms need food too in order to multiply. We also add brown sugar and mix it with animal faeces. It is important to ensure the antibiotics in the animal faeces do not exceed the standard parameters. Among the heavy metal elements, a highly cancerogenic substance is chromium. After a few years of cultivation in our field, the rice we produced was tested 37.04 times lower than the heavy metal indicator prescribed by the green food. Now you tell me if our food is safe or not.

A: (nodding)

B: It is thus very safe. If you have more questions please ask because we don't have much time now.

A: Ah no, I have basically all questions covered. I just wonder if it is possible that you can introduce a couple of farmers from your cooperative for such interviews in a later stage in case further data collection is needed. They can be also farmer representatives like you. Are there also farmers who are willing to participate in this in-depth interview with open questions?

B: Well (pondering for a few seconds), as for other farmers, they might have problems providing information like technical data and some concepts. I am confident that I answer better than them because I evaluate myself as a student who is willing to learn. Their answers must be less clearer than mine, so of course, if you ask me questions, I always try to answer. If not, I can ask my teachers to find time and educate me then I can answer you. However, during this time of year we are all very busy with production.

A: Ok, I see.

B: Yes.

A: Right, I understand, thank you for attending this interview! I think I have all data needed, but just in case I might have a bit more questions at a later stage I will turn to you again.

B: Oh, that's not a problem! I hope that after this interview you can transcribe it into words and if you will use any of these data for your paper, please send a digital format to me.

A: Yes, it goes like this: After data collection I will firstly transcribe it into words.

B: Process it properly.

A: Yes, after transcription I will add it to my paper, and then, hmm, I have not yet decided when the paper will be published, but this paper will use a case study that focuses on your cooperative and the interview regarding the non-chemical rice farming system. Therefore, the information I might use all comes from this interview, and after that ...

B: Well ...

A: Continue please.

B: Mr. Wu, I really hope you can promote the philosophy from our ancestors, i.e., the concept of harmony between man and nature, together with this farming system abroad. I hope more people can better communicate our Chinese philosophy so the people outside China can better understand it.

A: Yes.

B: It depends on you, as a young man, to communicate this culture.

A: (nodding)

B: Yes, this is the last thing I wanted to say.

A: Yes, of course. Thank you so much. I am really grateful that you can take the time to attend this interview as it is a busy season for agriculture. Actually, I should have gone to China and communicate with you face to face. However, due to the pandemic, I really cannot come back quickly because if I come back I will be firstly under quarantine for one month. Thus, I am thankful that you are able to attend the interview through internet. I would like to remind you that the information I will use in my paper all comes from our conversation and it will not involve any of your private information. I will only mention your cooperative, the Youjun village, and where your field is. This is one focus of my study. Just to let you know.

B: Okay.

A: So all the things you mentioned in the interview will be processed in confidentiality and for research use only.

B: This is absolutely necessary. If your paper is … I am thinking your paper can consider two aspects. The first one is to let others, no matter it is your professor, your classmate or your fellows in the university, know something about the Chinese traditional farming culture and our philosophy. Based on this, the ecological agriculture with the use of modern technology is able to develop steadily and go further.

A: (nodding)

B: Maybe I am wrong because I am not as educated as you so I cannot describe it in a good way. It is up to you how much information you want to use in your paper, ok.

A: Ok, thank you so much. I really appreciate it. Let's keep in touch and I shall let you do your work now since you are so busy.

B: Yes, that's good.

842 A: Ok keep in touch and good bye.

843 B: Yes, good bye.

Conferences
Policy Studies Organization hosts numerous conferences, including the Middle East Dialogue, Space Education and Strategic Applications, International Criminology Conference, Dupont Summit on Science, Technology and Environmental Policy, World Conference on Fraternalism, Freemasonry and History, AI – The Future of Education: Disruptive Teaching and Learning Models, Sport Management and Esport Conference, and the Internet Policy & Politics Conference. Recordings of these talks are available in the PSO Video Library.

Yearbook
The Policy Yearbook contains a detailed international listing of policy scholars with contact information, fields of specialization, research references, and an individual scholar's statements of research interests.

Curriculum Project
The Policy Studies Organization aims to provide resources for educators, policy makers, and community members, to promote the discussion and study of the various policies that affect our local and global society. Our curriculum project organizes PSO articles and other media by easily serachable themes.

For more information on these projects, access videos of past talks, and upcoming events, please visit us at:

ipsonet.org

Related Titles from Westphalia Press

The Limits of Moderation: Jimmy Carter and the Ironies of American Liberalism by Leo P. Ribuffo

The Limits of Moderation: Jimmy Carter and the Ironies of American Liberalism is not a finished product. And yet, even in this unfinished stage, this book is a close and careful history of a short yet transformative period in American political history, when big changes were afoot.

The Zelensky Method by Grant Farred

Locating Russian's war within a global context, The Zelensky Method is unsparing in its critique of those nations, who have refused to condemn Russia's invasion and are doing everything they can to prevent economic sanctions from being imposed on the Kremlin.

Sinking into the Honey Trap: The Case of the Israeli-Palestinian Conflict by Daniel Bar-Tal, Barbara Doron, Translator

Sinking into the Honey Trap by Daniel Bar-Tal discusses how politics led Israel to advancing the occupation, and of the deterioration of democracy and morality that accelerates the growth of an authoritarian regime with nationalism and religiosity.

Notes From Flyover Country: An Atypical Life & Career by Max J. Skidmore

In this remarkable book, Skidmore discusses his "atypical life and career," and includes work from his long life in academe. Essays deal with the principles and creation of constitutions, anti-government attitudes, the influence of language usage on politics, and church-state relations.

The Athenian Year Primer: Attic Time-Reckoning and the Julian Calendar
by Christopher Planeaux

The ability to translate ancient Athenian calendar references into precise Julian-Gregorian dates will not only assist Ancient Historians and Classicists to date numerous historical events with much greater accuracy but also aid epigraphists in the restorations of numerous Attic inscriptions.

Siddhartha: Life of the Buddha
by David L. Phillips,
contributions by Venerable Sitagu Sayadaw

Siddhartha: Life of the Buddha is an illustrated story for adults and children about the Buddha's birth, enlightenment and work for social justice. It includes illustrations from Pagan, Burma which are provided by Rev. Sitagu Sayadaw.

Growing Inequality: Bridging Complex Systems, Population Health, and Health Disparities
Editors: George A. Kaplan, Ana V. Diez Roux, Carl P. Simon, and Sandro Galea

Why is America's health is poorer than the health of other wealthy countries and why health inequities persist despite our efforts? In this book, researchers report on groundbreaking insights to simulate how these determinants come together to produce levels of population health and disparities and test new solutions.

Issues in Maritime Cyber Security
Edited by Dr. Joe DiRenzo III, Dr. Nicole K. Drumhiller, and Dr. Fred S. Roberts

The complexity of making MTS safe from cyber attack is daunting and the need for all stakeholders in both government (at all levels) and private industry to be involved in cyber security is more significant than ever as the use of the MTS continues to grow.

Female Emancipation and Masonic Membership: An Essential Collection
By Guillermo De Los Reyes Heredia

Female Emancipation and Masonic Membership: An Essential Combination is a collection of essays on Freemasonry and gender that promotes a transatlantic discussion of the study of the history of women and Freemasonry and their contribution in different countries.

Bunker Diplomacy: An Arab-American in the U.S. Foreign Service by Nabeel Khoury

After twenty-five years in the Foreign Service, Dr. Nabeel A. Khoury retired from the U.S. Department of State in 2013 with the rank of Minister Counselor. In his last overseas posting, Khoury served as deputy chief of mission at the U.S. embassy in Yemen (2004-2007).

Managing Challenges for the Flint Water Crisis Edited by Toyna E. Thornton, Andrew D. Williams, Katherine M. Simon, Jennifer F. Sklarew

This edited volume examines several public management and intergovernmental failures, with particular attention on social, political, and financial impacts. Understanding disaster meaning, even causality, is essential to the problem-solving process.

User-Centric Design by Dr. Diane Stottlemyer

User-centric strategy can improve by using tools to manage performance using specific techniques. User-centric design is based on and centered around the users. They are an essential part of the design process and should have a say in what they want and need from the application based on behavior and performance.

Masonic Myths and Legends by Pierre Mollier

Freemasonry is one of the few organizations whose teaching method is still based on symbols. It presents these symbols by inserting them into legends that are told to its members in initiation ceremonies. But its history itself has also given rise to a whole mythology.

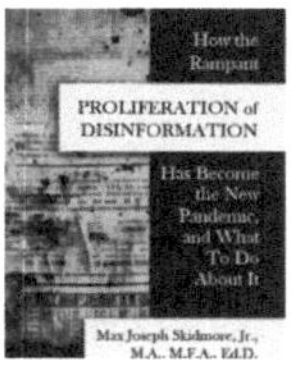

How the Rampant Proliferation of Disinformation has Become the New Pandemic by Max Joseph Skidmore Jr.

This work examines the causes of the overwhelming tidal wave of fake news, misinformation, disinformation, and propaganda, and the increase in information illiteracy and mistrust in higher education and traditional, vetted news outlets that make fact-checking a priority

Made in United States
Troutdale, OR
11/03/2026

56845239R00076